Journey into Geometries

Marta Sved

MAA
SPECTRUM

urney ínto Geometríes

Marta Sved

with Foreword by H. S. M. Coxeter
Illustrations by John Stillwell

THE MATHEMATICAL ASSOCIATION OF AMERICA

SPECTRUM SERIES

Published by
THE MATHEMATICAL ASSOCIATION OF AMERICA

———

In memory
of
Carl Moppert

Printed in the United States of America
Current printing (last digit):
10 9 8 7 6 5 4 3 2 1

Cover art by Christine Vadasz and Geraldine Searle

Foreword

T HIS *'Journey into Geometries'*, enlivened by conversations between Alice, Lewis Carroll, Dr. Whatif and two hostesses, may well be regarded as the geometric counterpart of *Kandelman's Krim* (Jonathan Cape, London, 1957) by Professor J. L. Synge, an Irish mathematician who was born in 1897. His five characters consist of a sky-blue Goddess, an Orc, a Kea, a Unicorn, and a Plumber, all discussing foundations of arithmetic. With a similar blend of mathematics and whimsy, Marta Sved gently leads her readers into areas of geometry which they had never properly understood before, very sensibly using straightforward notation, denoting points by capital letters (as Euclid himself did) and allowing the context to determine whether AB means a line, a length, a line-segment or a directed line-segment.

A conversational style like hers was employed earlier by C. L. Dodgson in his *"Euclid and his Modern Rivals* by Lewis Carroll" (Dover, New York, 1973). His TABLE III (pp. 37–38) contains five historical attempts to replace Euclid's Fifth Postulate by a more 'self-evident' statement. Later, in *Curiosa Mathematica* (3rd ed., London, 1890, p. 14), Dodgson added a sixth such statement:

'In every circle, the inscribed equilateral tetragon is greater than any one of the segments which lie outside it.'

It is tantalizing to reflect that, by denying any one of those six statements, he could have entered the Wonderland of Hyperbolic Geometry.

H. S. M. Coxeter

vii

Contents

Preface

EOMETRY is fun. Mathematicians know this, of course, but the point is often lost on laymen and students. One reason for this lies in the gap between mathematical discovery and exposition. Intuition and imagination spark off discovery somewhere in the deep, dark recesses of the mind. It is then lifted to the surface, checked and guided by strict rules of logic until crystallised into perfection, ready to be discharged by the peripheral processor of the brain. It is this very perfection which can be so disconcerting to the student who finds it cut and dried, removed from his own imagination and thinking processes. It is difficult for him to see what makes the real mathematician tick, to perceive the double personality hiding in his mind: not a Dr. Jekyll and a Mr. Hyde, but more appropriately a Rev. Dodgson and a Lewis Carroll.

This little book attempts to bring modern geometry nearer to the imagination of the reader. We must apologise to the late Rev. Charles Dodgson (alias Lewis Carroll) and Mrs. Hargrave (née Alice Liddell) for sublimating their historical figures into fictitious characters to serve as vehicles for the long journey into modern geometry.

This book, though not a formal text, is first and foremost about geometry. It is neither comprehensive, nor can it claim to go very deep into the chosen topics, yet hopefully, it may initiate a spark to light the way into further progress. There is no lack of excellent text books for the follow up. The central topic in this book is non-Euclidean geometry. The approach to it is made via the *Poincaré* model, or rather a variation of it used by the late Carl Moppert in university lectures given in Australia and Switzerland (Monash, Basel). The theory of Steiner families of circles and inversion leading up to the models are discussed in detail. After stating that this is neither a formal geometry text, nor a story book, two questions should be answered: How should the book be read and who should read it?

The background knowledge assumed on the part of the reader is not more than the basic high school course in elementary geometry and elementary algebra. This does not mean that no mathematical demands are made on the reader. Although the treatment is informal, the purpose of each chapter is to communicate geometrical content.

The book can be read on two levels. A first reading should give a glimpse of modern geometry to the uninitiated while providing amusement to the expert. A second reading is required for serious study, for which Part II was added. To help the exposition, a set of exercises and problems is attached to each chapter. These problems form an integral part of the text.

The completeness of the exposition depends on them, more so than in a conventional text. They are there also to promote active participation on the part of the reader, as in other text books. Since the results are essential, solutions to the problems are supplied at the end of the book.

It is clear from all that precedes that the book is directed as background reading to modern geometries, at students and teachers of mathematics at tertiary or advanced secondary level, also at interested laymen, even if their school mathematics has gone somewhat rusty; not much of it has to be recalled: concepts of congruence and similarity and elementary properties of the circle, easily found in simple school texts.

Is it a sacrilege to introduce madness into the method and mix mathematics with mundane irrelevancies, irreverent doggerels, and puns? We hope that it is not so. We believe that a light-hearted approach, far from impairing the process of learning, could stimulate thinking and add depth to understanding.

Acknowledgements

M Y thanks go first and foremost to the late Dr. Carl Moppert who inspired this work with his lecture course on geometry to undergraduates at Monash University. I am grateful to Professor Eric Barnes of the University of Adelaide who helped me to revise some details and correct a few errors in the text and to Dr. John Stillwell, also of Monash University, who turned my rough sketches into delightful illustrations and neat diagrams.

I thank Professor Coxeter for his encouragement and suggestions for some improvements.

I also thank Susan Renshaw, Judi Hurley, and Mary Mattaliano for typing successive versions of the book.

My thanks also go to Beverly Ruedi for setting out and assembling the work with care and clear judgement.

Last, but not least, I would like to thank my husband, George, for his support and help.

Reader's Guide to the Journey

HE book consists of two parts. Part I is the Journey.

In Part 2 the goods acquired on the Journey are unpacked and put into order. Thus Part 2 is more formal. It contains the axiom-systems of the various geometries, the solutions of the problems presented to Alice and a bibliography for further reading.

The problems which appear in Part 1 at the end of each chapter complement the content of the chapters, in which the ideas are communicated in an informal manner. The problems vary greatly: some of them are exercises, others are quite difficult. In Part 2 the solutions are given in detail, but the understanding of these will be more complete if you try to solve them first. Even if your efforts are unsuccessful, you gain a deeper insight than if you go straight to reading the solutions.

On Notations: Conventions in geometry text books vary greatly. I did not want to clutter the text with too many symbols. Admitting that the symbol AB could mean various things: the line AB, the line-segment AB, the length of the segment AB, the path beginning at A and ending at B, I did not use a different notation for each meaning, because each time this should be clear from the context.

Angles are marked either by a Greek letter, α, β, etc., or by the symbol $\angle ABC$ where B means the vertex, A and C are points on each arm.

The symbols $\cong$ and $\sim$ for congruence and similarity are standard and should be familiar from schoolwork.

Have a safe and happy journey.

Part I

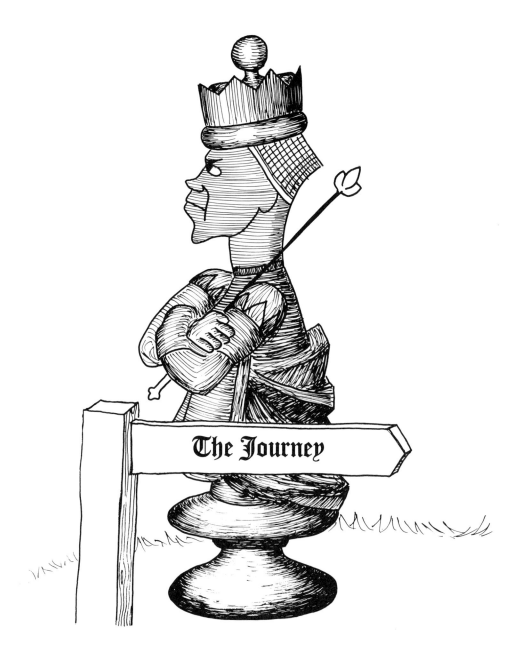

The Journey

Introduction

ALICE was standing in front of her mirror. It was mid-afternoon on a summer day and she felt hot and tired in her tightly laced stays.

"Since the time I left school," she reflected, reflected ... Alice heard herself speaking, or did she hear strange voices? "Of course, reflecting is just the thing to do when you are in front of a mirror. Oh, Alice, be careful, those days through the looking glass are long past. I was a child then, now I am past finishing school."

"Hold yourself straight Alice, straight, straight, and walk away from that glass!" It was the voice of Miss Prim, that strict governess, or was it the Red Queen?

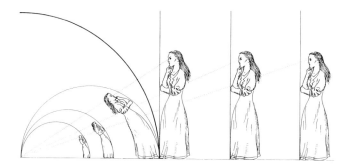

"Oh, oh, I am bending backwards! And in a perfect circular arc, too. I could never do it before! My head will soon reach the ground."

"Your head will never reach the ground, you poor unliberated Victorian creature," boomed the voice. No, it was not Miss Prim. It was not the Red Queen either.

"Of course, you've never heard of *inversion*."

"Inversion??"

"Of course, you have not heard of it. You hardly know *any* geometry!"

"But I do know my geometry. I do! I do! I know my Euclid. I know all the *postures*."

"Say *that* to Miss Prim. You mean postulates!"

"Oh, yes, postulates, and prepositions too!"

"Do you know anything about non-Euclid? Is your geometry liberated?"

Alice leaned against the French window. She opened her eyes, breathing hard. "I must see Uncle Lewis Carroll. I am so confused."

A few minutes later she was in Lewis Carroll's study. It was dark and cool and lined with volumes and volumes of mathematics. She tried to recall her thoughts or her dream or whatever it was.

"You are a deep sinker, child," said Lewis Carroll.

"You mean, deep thinker."

"I mean, deep sinker. What about that time when you went down the rabbit hole? You nearly fell through the Earth! Why, had I not saved you, you would have gone all the way to meet the *antipathies*. (That's what you called them.) But, my dear, you could not have stopped there."

"Why not?"

"You would have fallen *back*.

> The rabbit hole's not hard to enter,
> But it runs right through the centre,
> Through hemispheres South and North
> Alice sallies back and forth,
> Like a pendulum, like a swing
> Like a tuning fork, like a spring;
> Her predicament is chronic,
> She moves in a *simple harmonic*."

"Oh, Uncle Lewis Carroll, I wish you talked *sense* just for once! I was not speaking about the rabbit hole. I was speaking about the looking glass and about *geometry*."

"*Geometries*"

This was another voice. Alice noticed another figure amongst the dark shadows. Was it the Mad Hatter without his hat?

"Alice, have you met Dr. Whatif?" asked Lewis Carroll.

"Pleased to meet you, Dr. Whatif. You have an unusual name, Sir. Do you come from Russia or thereabouts?" Alice sounded bewildered.

"I prefer to travel in *time*. As a matter of fact, I have come on a trip from the twentieth century. I enjoy such *incursions*."

"You mean, excursions," Lewis Carroll tried to correct him.

"Incursions. I feel that all of you here need a little mind broadening. Leaving other things aside, take, for example, geometries ..."

"Geometries!" called Alice. "Why, Dr. Whatif, you remind me of a conversation I had with the Red Queen. I met her when I went through the looking glass. It was very long ago."

"The Red Queen?" This time Dr. Whatif was puzzled.

"She told me that they have a lot of Tuesdays, all at once. You, in the twentieth century have a lot of geometries all at once. Is it because you are all red?"

"Neither all red, nor all royalty. But we have several geometries. Speaking truthfully, they are not all of our making. You people in the nineteenth century started their development. Take Bolyai."

"And Lobachevsky, Gauss, and Riemann, and all those other fellows on the Continent," continued Lewis Carroll. "They fail to convince me. Though no flaw is visible in the arguments of the modern rivals of Euclid, the Truth of nature sides with Euclid."

Dr. Whatif smiled, turned to Alice, and asked, "Young lady, how is your Greek?"

"Greek? Why, we girls here in Oxford ...."

"Are not sufficiently liberated to have all your liberated hours filled with Greek and Latin. So do you even know the meaning of the word 'geometry'?"

"Yes, I do," said Alice. "'Measuring the earth', isn't it?"

"Splendid, splendid, and how do you measure the earth?"

"Why, with measuring rods."

"Like the one here." A measuring rod appeared suddenly in Dr. Whatif's hand. "Will you measure the edge of the table?"

"Why, this is not a measuring rod, this is a caterpillar!" exclaimed Alice with horror as she tried to move the rod along the table. She threw it down abruptly.

"Don't talk nonsense, my dear young lady," said Lewis Carroll. "It is just a ruler!"

"It *was* a caterpillar when I moved it. It shrank."

"But it is a *rigid* ruler. Can't you see it for yourself?"

"Are you sure that she talks nonsense? She may have *extraordinarily* fine eyesight! Maybe the ruler *does* shrink!" said Dr. Whatif.

"It is rigid!" countered Lewis Carroll.

For a moment the two dons looked at each other, then both called out at once,

"Yes, it *can* change."

"No, it can't change when you move it."

"I challenge you, can you prove it?"

Excited, they went on chanting, "Can you prove it, when you move it? Can you prove it, when you move it? ..."

Bewildered, Alice looked from one to the other. They stopped abruptly.

Whatif turned to Alice. "All right. Let us *pretend* that this rod is *not* a caterpillar. I give you as many days as you like for measuring out with it a length of 1000 miles in a *straight line*."

Quickly, Alice began to move the rod along the table, keeping her *extraordinarily* sharp eyes sufficiently far from it so as not to see the contraction of the caterpillar. She saw soon that she would not get very far moving it on the table.

"Use the floor, Alice. Then go out with it to the garden."

Alice looked at the two of them and saw that this time both were grinning in agreement. She suddenly saw the trick. "I cannot do it! There is *no* straight line on the Earth 1000 miles long. You both *know* the earth is round."

Dr. Whatif laughed. "So much for your plane geometry! 'Measuring the earth,' by moving your rod which is *really* a caterpillar in a plane which exists only in your *mind.* How can you say, my friend Lewis Carroll, that the geometry of Euclid is truth, but those of Bolyai, Lobachevsky, and the other continental fellows lead to nonsense?"

More perplexed than ever, Alice turned to her old friend, Lewis Carroll. "Please, I am not a child any more. Will you tell me what *is* that talk about Bolyai and the rest of them?"

"All right, Alice. Do you know what parallel lines are?" asked Dr. Whatif.

"Of course I do. They are lines in the plane which never meet."

"Now, then. You have also learned the fifth postulate of Euclid?"

Alice drew a breath and started reciting, "If a straight line falling on two straight lines . . . ."

"I see that you know it. Putting it simply, the postulate states that through any point not lying on a given straight line we can draw *exactly one* line parallel to the given line."

"It is *obvious* to me," cried out Alice.

"Well, to some people it was not obvious and so they tried to prove it, but no one succeeded," remarked Lewis Carroll.

Dr. Whatif continued, "So it occurred to those fellows we were speaking about that they could make different assumptions: *either*

(1) through any point not lying on a given straight line we can draw *no* line parallel to the given line, *or*

(2) through any point not lying on a given straight line there are at least two distinct lines parallel to the given line.

A geometry, different from that of Euclid, can be based on each of these two postulates."

Lewis Carroll called out, "Yes, but it's all purely fiction!"

"Still, there is no contradiction," countered Dr. Whatif.

"They will break out again in song," thought Alice with alarm. However, to her great relief Dr. Whatif continued in prose and he spoke quite seriously.

"My mathematician friends in the twentieth century are more modest in their aims than you are. They have given up chasing the 'truth.' They do not even try to define the simplest terms that we are speaking about. They know the definition game would never end."

"But I have learned *all* the definitions of Euclid," said Alice eagerly and she immediately began to recite,

"1. A *point* is that which has no part.

2. A *line* is length without breadth.

3. The extremities of a line are points.

4. A *straight line* is a line which lies evenly with the points on itself . . . ."

She wanted to continue, but Dr. Whatif stopped her.

"Why don't you tell me then what the words 'part,' 'length,' 'breadth' mean and why don't you explain how one should 'lie evenly with the points on itself'?"

Alice looked aghast.

Dr. Whatif continued, "So you see, my friends in the twentieth century occupy themselves with *'what if'* mathematics."

"Oh, you are so famous that they named their mathematics after you?"

"Well, it is the other way round. They named me after their mathematics. We like to do things the other way round. Instead of things which are thought to exist, but whose properties we do not know for certain, we prefer to speak about things of which the properties are certain, not worrying whether they exist."

"Exist? What does it mean anyway?" asked Alice.

"A good question! So we speak about points and lines and make up statements, called *axioms*, just like Euclid, such as these:

1. Through any two distinct points there is exactly one line.

2. Every line has at least two distinct points.

3. Not all points are on one line."

"What then are the *lines* and *points*?" asked Alice.

"A good question. We leave it at that."

"But if you do not even know what the points and lines are, how do you know that your statements are true?"

"We do not care. The axioms are 'if' statements. The geometry that follows from them is 'what if' geometry."

"All right," said Alice, "suppose that we have for *points*:

chocolate
peppermint
nuts

and for *lines*:

peppermint chocolate
peppermint nuts
chocolate nuts.

Is this a geometry?"

"Yes, a very tasty one. All axioms are satisfied."

"Tasty, but not a very *rich* geometry," countered Lewis Carroll. "Not too many points and lines in it."

"We can alter the axioms to *force* in a few more points. Say that we keep axioms 1 and 3 and alter axiom 2 so that we have: *2. There are exactly 3 points on each line.*

Add for good measure:

4. There exists one line.

(So that we know we are speaking about *something*.) And, finally:

5. There is at least one point on any two distinct lines.

Out of all this we can construct a *gemmetry* and present it to Alice."

"A gemmetry?" asked Alice.

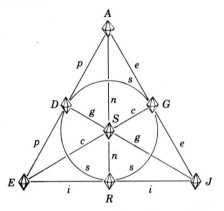

"Yes. Our points will be gemstones and our lines will be chains containing the gemstones. It will be like this: the points will be amethyst (*A*), diamond (*D*), emerald (*E*), garnet (*G*), jade (*J*), ruby (*R*), sapphire (*S*). The chains will be made out of copper (*c*), enamel (*e*), gold (*g*), ivory (*i*), nickel (*n*), platinum (*p*), and silver (*s*), each containing exactly three gemstones. What is more, you can prove, if you wish, that with only these axioms we can make up this sort of seven point geomtry."

"But," said Alice, "couldn't we *pretend* that the *chains* are the *points* and the *gems* are the *lines*?"

"That is the *dual* geometry, I give you your due," said Dr. Whatif, "but it is still a seven point Fano geometry with the same structure."

Lewis Carroll was losing his patience,

> "When everything is said and done,
> Fano's only finite fun."

"You would be surprised, how many interesting things you can do with it," answered Dr. Whatif.

"Still," said Alice, "this is nothing like the good geometry we had at school with so many circles and triangles and squares and hypotenuses and problems and theorems and proofs. I just *loved* it."

"I only wanted to illustrate how axioms can be built up into a geometry. I do not deny that the Euclidean plane is *rich*, but with different axioms you can make *other* rich and varied geometries, and the whole lot together is more varied and exciting than you ever *dreamt* of."

"One moment," said Lewis Carroll, "you surely cannot just make up sets of axioms and then hope for the best!"

"Certainly not," said Dr. Whatif and turning to Alice he added,

> "With axioms, my dear, you need a gentle touch,
> They should not say too little, they should not say too much,
> And on one point above all, we have to be insistent,
> Though axioms need not be 'true,' their *set* must be consistent."

"Very neat," said Alice, "but can you make it clearer?"

"The most important thing is *consistency*. There should not be any contradiction in the geometrical theorems we deduce from the axioms. The other things are desirable, too. The axioms should be independent. If one of them can be deduced from the others, then it is not an axiom any more, but a theorem. If the axioms say too little, like the three axioms from which you obtained your sweet and tasty geometry, then there are too many different geometries which satisfy the requirements. For example, the Fano geometry, the Euclidean geometry, and the Bolyai-Lobachevsky geometries *all* satisfy those axioms."

"How can you make sure that the axioms are consistent, that there is *no* contradiction?" asked Alice.

"It is difficult to be sure. With Euclidean geometry we have many centuries of experience. In the 20th century, its foundations have been tidied up a little and we find no contradictions in it. Moreover, we all feel at home in it, don't we?"

"Yes, we do," said Lewis Carroll and Alice at once.

"So, when we want to test some *non-Euclidean* axioms, we make use of Euclidean geometry to test them."

"How can that be done?" asked Alice.

"I can't tell you everything at once. I can only ask, are you willing to take a journey with me? We will make an exploring trip into the inversive plane. It is really still Euclidean geometry. Then we can proceed further. Alice, is your geometry really good?"

Alice answered,
"I know all the definitions,
Postulates and propositions,
And with maximum decorum
Crossed the great 'pons asinorum'."

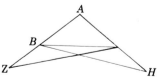

Here, Dr. Whatif looked a little puzzled. Lewis Carroll was quick to explain, "Pons asinorum is the 'bridge of the asses.' We gave this name to the theorem of Euclid which states that the base-angles of an isosceles triangle are equal. Euclid used a rather involved diagram to prove the theorem. It looks a little bit like a bridge, and students who could not recall Euclid's proof were silly asses."

"Once you struggled through the bridge, the road was still not easy," added Alice.

She continued, "I had a little conversation with the Red Queen when I met her. She asked me to show her the road to geometry.

'To geometry, madam, there is no royal road,' said I.

'We are not amused,' answered the Queen."

"I hope," said Dr. Whatif, "that *you* will be amused as we make our journey to geometry-wonderland."

Chapter 1

Going Round
in Circles

HEN Alice arrived at the meeting place for her first journey, she was happy to find Lewis Carroll alone, without Dr. Whatif.

"Whatif has gone out to shop for souvenirs. He says that Victoriana is in great demand in the twentieth century."

"What about our journey?"

"You can safely take me as your guide. Today we will not stray for one moment from the realm of Euclidean geometry."

"Oh, that is good! So much more familiar."

"Yes, we will be moving in a familiar circle, or more exactly, in *families of circles*."

"What???"

"To begin with, I take it that you feel really at home in the circle."

"Oh, yes. I always had fun with those riders, where you had to spot equal angles, like these." She quickly chalked a diagram on the slate.

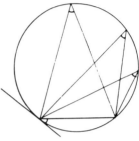

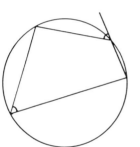

"Or like these, spotting external angles of cyclic quadrilaterals.

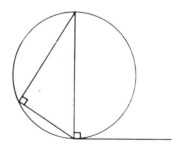

In particular, I liked to find the right angles in the figure, like these."

These figures prompted a dialogue between Lewis Carroll and Alice,

"So you are an expert angler."
"Though some problems may be curly."
"Shame, you can't become a wrangler[1]"
"Just because I am a girl?"
"Born a hundred years too early!"

---

[1]Wrangler is the name given to a person gaining first class honours in the mathematical tripos examination in Cambridge.

A new drawing appeared now on the slate. Alice looked at it.

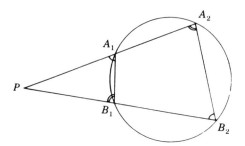

"Can you *prove* that $PA_1 \cdot PA_2 = PB_1 \cdot PB_2$?" asked Lewis Carroll.

"Let me see. Oh, yes. There are those *similar* triangles there, $\triangle PA_1B_1$ and $\triangle PB_2A_2$, so

$$\frac{PA_1}{PB_2} = \frac{PB_1}{PA_2}$$

or $PA_1 \cdot PA_2 = PB_1 \cdot PB_2$"

"Have you ever heard the name of the product $PA_1 \cdot PA_2$, or $PB_1 \cdot PB_2$ for that matter?" asked Lewis Carroll.

"No," admitted Alice. "Why does it have a special name?"

"The product is important, because if the line through P *sweeps through the circle*, the products $PA_1 \cdot PA_2$, $PB_1 \cdot PB_2$, and so on, remain the same. The product only depends on the positions of the point P and the circle $\mathbf{C}$."

"I spy something," said Alice triumphantly. "The two intersection points can come together in a single point T, so that PT is a tangent. Then

$$PT^2 = PA_1 \cdot PA_2."$$

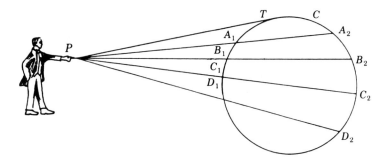

"You are right," said Lewis Carroll. "A little calculation will also show this. We have seen that A_1 can be chosen anywhere on the circle. If PA_1 meets the circle in A_2 the product $PA_1 \cdot PA_2$ is the same. So let us choose A_1 and A_2 along PO, where O is the centre of the circle. If you denote the distance of P from the centre

O by d and the radius of the circle $\mathbf{C}$ by r, then

$$PA_1 \cdot PA_2 = (d+r)(d-r) = d^2 - r^2.\text{''}$$

Alice exclaimed, "Then by Pythagoras,

$$PT^2 = PO^2 - OT^2 = d^2 - r^2 = PA_1 \cdot PA_2.\text{''}$$

"We will call the expression *the power of P with respect to the circle $\mathbf{C}$*," announced Lewis Carroll.

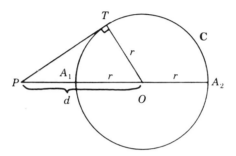

"Quite a mouthful," said Alice. "What happens if P is *inside* the circle or on the circle?"

"A good question. In that case $d^2 - r^2$ is negative or zero and, of course, there is no tangent from P to the circle, so $d^2 - r^2$ cannot be regarded as the square of the length of the tangent. In this case, the power of P with respect to $\mathbf{C}$ is zero or negative."

"I see it all now," said Alice. "P can only have a great power with *respect* to the circle if it respects it! From a respectful distance its power is great, but if it gets cheeky and gets nearer, its power decreases rapidly."

"This is the idea," said Lewis Carroll. "Let us put it all in a formula for future reference, $P(\mathbf{C}) = d^2 - r^2$,

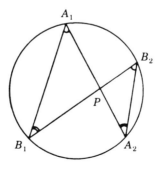

that is, the power (P for point and P for power) of the point P with respect to the circle $\mathbf{C}$ only depends on d, the distance of the point P from the centre of $\mathbf{C}$ and the radius r of $\mathbf{C}$."

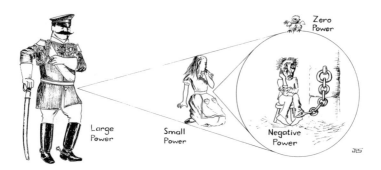

"You told me that we are going to meet *families* of *circles*," said Alice.

"That is just what we are going to do now. May I introduce you to this nice family?"

Alice looked at the slate, where a bunch of circles, of varying sizes appeared. "What makes them a family?"

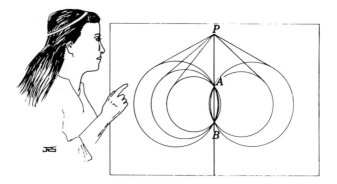

"They all intersect at the points A and B. The line AB is their common chord."

"So their centres all lie on the same straight line," said Alice.

"You will have to prove this in the evening, together with a few other problems we will take back from this trip. Can you find any other remarkable things when you climb up on the line AB?"

Alice pointed at the point P. "If I go to P, then my power with respect to all the circles is *the same*! $P(\mathbf{C}_1) = P(\mathbf{C}_2) = P(\mathbf{C}_3) \cdots$."

"How do you know this?"

"Well, every time it is $PA \cdot PB$."

"Did you strike a lucky point P on the line?"

"Of course not! This would hold for any point P as long as I stay on the power-line AB!"

"Power-line? That would be quite a good name for it! Only we do not want to

get mixed up with the electricity supplies of Dr. Whatif and his friends. So we call
AB

> *the radical axis of the family of the coaxial circles* $\mathbf{C}_1, \mathbf{C}_2, \dots$."

"Speaking about Dr. Whatif, what if the circles don't intersect in two points?"

"That can happen."

Alice looked at the new circles just touching at T. "I can see here that the common tangent at T is, ... What do you call it?"

"The radical axis."

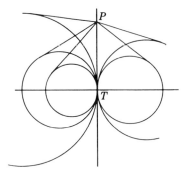

"But I really did not mean this case. I wanted to ask you a question. Is it possible to coax some circles into a coaxial family if they *never* meet?"

"What you are really asking is, can *two circles that do not intersect* have a radical axis? Are there any points having equal power with respect to the two circles? Look at the two circles $\mathbf{C}_1$ and $\mathbf{C}_2$ which do not intersect. Take a walk along the line joining their centres from A to B, that is, from circle $\mathbf{C}_1$ to circle $\mathbf{C}_2$.

The point A has some power with respect to $\mathbf{C}_2$, but zero power with respect to $\mathbf{C}_1$. What happens when you reach B?"

"The power with respect to $\mathbf{C}_2$ dwindles to zero, but we pick up some power with respect to $\mathbf{C}_1$."

"So somewhere on the way you should find a point E such that

$$E(\mathbf{C}_1) = E(\mathbf{C}_2).$$

In fact, before you go to sleep this evening, you can calculate the position of E."

"And I bet," said Alice, "that the power-line ... er ... radical axis is the perpendicular to AB, erected at E."

"You do not bet," said Lewis Carroll disapprovingly, "you guess, you conjecture, then you try to prove it—or let future generations of mathematicians try to prove or disprove your conjecture."

"We do not need future generations for this one," said Alice (who really was a quick thinker). "I bet ... oh ... I conjecture that old Pythagoras will do the job."

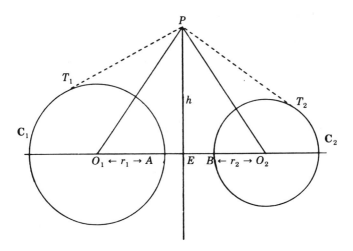

"So he will," said Lewis Carroll. "In fact, can you show that if P is on the perpendicular to AB at E and $PE = h$, then

$$P(\mathbf{C}_1) = E(\mathbf{C}_1) + h^2?"$$

Alice thought for a short while, then quickly wrote, "Let r_1 be the radius of $\mathbf{C}_1$.

$$\begin{aligned}
P(\mathbf{C}_1) = PO_1^2 - r_1^2 &= \left(EO_1^2 + EP^2\right) - r_1^2 \\
&= h^2 + \left(EO_1^2 - r_1^2\right) \\
&= h^2 + E(\mathbf{C}_1). \quad \text{Q.E.D.!!"}
\end{aligned}$$

"All right, you convince me without the two exclamation marks. But you have not proved yet that the line PE is the radical axis of $\mathbf{C}_1$ and $\mathbf{C}_2$."

"That is easy from here. Since

$$E(\mathbf{C}_1) = E(\mathbf{C}_2)$$

and

$$P(\mathbf{C}_1) = E(\mathbf{C}_1) + h^2$$

and in the same way

$$P(\mathbf{C}_2) = E(\mathbf{C}_2) + h^2$$

it is clear that

$$P(\mathbf{C}_1) = P(\mathbf{C}_2)."$$

"What is more," added Lewis Carroll, "you can also argue conversely. If you know that for some point P

$$P(\mathbf{C}_1) = P(\mathbf{C}_2)$$

and the foot of the perpendicular from P to O_1O_2 is F, then it is also true that

$$F(\mathbf{C}_1) = F(\mathbf{C}_2),$$

hence F must be the point E, since there is exactly one point on the line O_1O_2 with that property. So *all* the points of the radical axis have equal powers with respect to $\mathbf{C}_1$ and $\mathbf{C}_2$ and, conversely, all the points with equal powers with respect to $\mathbf{C}_1$ and $\mathbf{C}_2$ must be on the radical axis. Any questions?"

"Yes. Can you *coax* other circles to join the family of $\mathbf{C}_1$ and $\mathbf{C}_2$?"

"Any number you like. Just select a centre O_3."

"Anywhere?" asked Alice.

"You should know better than that! The line EP is perpendicular to O_1O_2."

"So it must be perpendicular to O_1O_3 too."

"So?"

"So O_1O_3 is the same line as O_1O_2. You can only draw one perpendicular from O_1 to a line. You see, I know my Euclid. O_3 is on the line O_1O_2. Can I select O_3 anywhere on the line?"

"You will answer that question yourself," stated Lewis Carroll. "Suppose that I select a good point O_3 for you. Can you construct the circle $\mathbf{C}_3$?"

"Well ...,

$$E(\mathbf{C}_1) = E(\mathbf{C}_3).$$

So the length of the tangent from E to $\mathbf{C}_1$ is the same as from E to $\mathbf{C}_3$ ... ."

Lewis Carroll helped with a drawing and a hint. "The situation is that the tangents to the *given* circle $\mathbf{C}_1$ and the *wanted* circle $\mathbf{C}_3$ are ET_1 and ET_3 and they are equal."

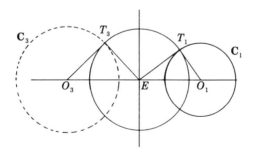

"So T_1 and T_3 *are on a circle around* E. I must find first T_1, then T_3."

"That should be easy for somebody who knows her school geometry as well as you do. You know how to draw a tangent from a point to a circle."

"I do. So I draw one from E to $\mathbf{C}_1$ to find T_1, then I draw that circle around E, then ... ." She chewed her finger.

"How about drawing a tangent O_3T_3 from O_3 to that circle about E?"

"But it is ET_3 that is a tangent to the circle $\mathbf{C}_3$!"

"Of course, but you have the situation of orthogonal circles."

"*Orthogonal* circles? What are they?"

"Two circles which intersect at *right angles*. Their tangents at the point of intersection are at right angles. So the tangent of one circle goes right through the centre of the other and this is mutual."

Alice gave a little shiver. "It looks to me like a cruel confrontation. Each circle aims its dagger right at the heart of another!"

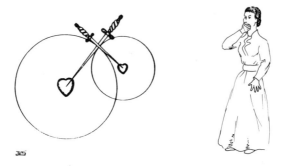

JeS

"You are still suffering from bad dreams, child. I hardly dare to tell you the story of the orthogonal families of circles. Anyway, once upon a time there was a happy coaxial family enjoying their togetherness at the points A and B. We will call them the elliptic family."

"You have seen earlier that any point of the radical axis has the same power with respect to all the circles of the elliptic family, so the tangents PT_1, PT_2, PT_3 drawn to the circles C_1, C_2, C_3 are all equal, hence $T_1, T_2, T_3, \ldots$ are on a circle about P. Now you can draw such a circle about every point of the radical axis AB, provided that the point is not between A and B. Every one of these circles is orthogonal to every one of the circles of the elliptic family. So I have introduced you to a new family, the *hyperbolic family*."

"And *each* member of the *hyperbolic* family confronts *each* member of the *elliptic family orthogonally*, just like the Montagues and the Capulets?" asked Alice in a horrified voice.

"But you will not find a Romeo and Juliet amongst them, so just shake off your nightmarish romantic notions and try to find the radical axis of the hyperbolic family."

Lewis Carroll took on a stern, no-nonsense look, which Alice found quite unusual. She made an effort and looked hard at the two families, "If I pick *any* circle in the elliptic family, then I can look at *its radii* as being *tangents to* circles of *the hyperbolic family*, so its centre has equal power with respect to the circles of the hyperbolic family."

"So?"

"So, the radical axis of the hyperbolic family is the line containing the centres of the elliptic family, just as the radical axis of the elliptic family holds the centres of the hyperbolic family."

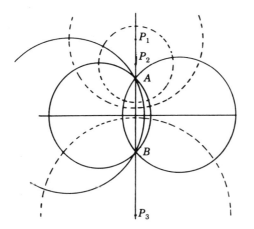

Elliptic family ——

Hyperbolic family ----

At this moment the door opened and Dr. Whatif appeared huffing and puffing, "I have been going round circles all day."

"So have we," said Alice gaily. She was given to quick changes of mood and her success in finding the radical axes of the orthogonal families made her forget all her notions about murderous Montagues and Capulets.

"Will you join us?" she asked Dr. Whatif. So the three of them set out on a gay romp of the orthogonal circles.

Suddenly Dr. Whatif grabbed at point *B*.

"Whatif!" cried out Lewis Carroll.

"What if," asked Whatif, "I remove this point to infinity?" And with these words, point B and Dr. Whatif disappeared. Then something strange happened. The circles of both families also made a move. The elliptic family hung on to point A, which made every circle grow until they did not look like circles at all (at least that is what Alice thought). The hyperbolic family, anxious to keep orthogonal, also moved and when the movement stopped ("Did it really stop, or is it too hard to detect that it still moves?" thought Alice), the emerging picture was like this.

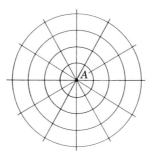

"I wonder," asked Alice, "if Dr. Whatif has reached infinity with his point B?"

"Well, the elliptic family turned into straight lines and the hyperbolic ones became concentric! Now one rule of coaxial families is that their members cannot be concentric!"

Suddenly, from a direction *opposite* to the one in which he had disappeared, Dr. Whatif reappeared again. Alice gasped, "Which way is infinity?"

Dr. Whatif laughed and grabbed point A and quickly vanished again with it. Alice rubbed her eyes. It was all so quick that she did not even see which way he took point A. She only saw that the circles moved again.

"I don't even know which were the elliptic and which were the hyperbolic family," said Alice.

"It does not matter much by now. They *both* turned into circles with centres in infinity! Let us go now and have a rest before our next journey. However, I see that you will need some time by yourself for thinking over things."

He handed Alice a sheet of problems. "After each journey you can take home your problem packet. Here are the ones for today."

Problems and Exercises 1

1. The power of a point P with respect to the circle $\mathbf{C}$ is defined as

$$P(\mathbf{C}) = d^2 - r^2 \qquad (1)$$

if P is outside the circle $\mathbf{C}$ and d is the distance of P from the centre of $\mathbf{C}$ and r is the radius of $\mathbf{C}$.

If P is *inside* $\mathbf{C}$ or on $\mathbf{C}$, then $P(\mathbf{C})$ is still defined by (1), but in these cases there is no tangent line from P to the circle and $P(\mathbf{C})$ is negative or zero. Show, however, that $|P(\mathbf{C})|$, the absolute value of $P(\mathbf{C})$, is equal to the product $PA_1 \cdot PA_2$, where A_1 and A_2 are two points on the circle such that the chord $A_1 A_2$ passes through P.

2. Prove that if a circle is orthogonal to two given circles, then its centre lies on the radical axis of the two circles. Show also that conversely: if a circle is known to be orthogonal to one of the circles and its centre lies on the radical axis of the two circles, then it is orthogonal also to the other circle.

3. It has been shown that the *radical axis of two nonintersecting circles* is determined by its intersection with the line joining the centres of the two circles. Let $\mathbf{C}_1$ and $\mathbf{C}_2$ be two circles having radii r_1 and r_2 respectively, the distance of their centres O_1 and O_2 being s, where

$$s > r_1 + r_2.$$

A is a point on the line $O_1 O_2$ such that

$$A(\mathbf{C}_1) = A(\mathbf{C}_2).$$

Prove that the distances $O_1 A$ and $O_2 A$ are

$$\frac{s^2 + r_1^2 - r_2^2}{2s} \qquad \text{and} \qquad \frac{s^2 + r_2^2 - r_1^2}{2s} \qquad \text{respectively.}$$

Find similar expressions for O_1A and O_2A when one circle lies *inside* the other.

4.

i. Show that the centres of the circles of an *elliptic family* (i.e., a set of circles with a common chord AB) lie on the same straight line. (As shown, this line is the radical axis of the hyperbolic family of the circles orthogonal to all the circles of the elliptic family.)

ii. $\mathbf{C}_1, \mathbf{C}_2, \mathbf{C}_3$ are three circles such that their centres do not lie on the same straight line. Let ℓ_1, ℓ_2, ℓ_3 be the radical axes of the pairs $\mathbf{C}_2\mathbf{C}_3, \mathbf{C}_3\mathbf{C}_1$, and $\mathbf{C}_1\mathbf{C}_2$ respectively. Show that ℓ_1, ℓ_2, and ℓ_3 are concurrent. What can you say

    a. if the centres of $\mathbf{C}_1, \mathbf{C}_2, \mathbf{C}_3$ are on the same straight line;

    b. if two centres coincide?

5. Find an easy construction for the radical axis of two nonintersecting circles. *Hint.* Use the result of 4.

6. $\mathcal{E}$ is an elliptic family of circles with common chord AB, of length a. $\mathcal{H}$ is the hyperbolic family of circles orthogonal to circles of $\mathcal{E}$. Find the *smallest* circle for each family, i.e., determine the radius and the position of the centre for each. Can you find a *largest* circle in either family?

7. Using the notations in 6, let r be the radius of a circle $\mathbf{C}$ belonging to $\mathcal{H}$. Assume that O, the centre of $\mathbf{C}$ is nearer to A than to B. Let x be the distance OA. Show that

$$x \le \frac{r^2}{a}.$$

Suppose that the point B is moved along the line AB, so that the distance AB tends to infinity and at the same time r, the radius of the circle $\mathbf{C}$, is kept fixed. What happens to O, the centre of $\mathbf{C}$?

8. P and A are two distinct points such that A is on the circumference of a circle $\mathbf{C}$. Construct a circle through A and P, orthogonal to $\mathbf{C}$. Is the construction always possible?

9. P is a point inside the circle $\mathbf{C}$, distinct from O, the centre of the circle. Show that all circles orthogonal to $\mathbf{C}$ and going through P intersect at a point P', where P' is on the line OP and

$$OP \cdot OP' = r^2$$

where r is the radius of $\mathbf{C}$. What happens if P coincides with O?

10. Let A and B be two points inside the circle $\mathbf{C}$, distinct from the centre. Construct a circle through A and B orthogonal to $\mathbf{C}$.
Hint. Problem 9.

11. C_1, C_2, C_3 are three circles such that their centres are not on the same straight line. Find a circle orthogonal to all three circles.

12. Through a given point P construct a circle orthogonal to two given circles C_1 and C_2. Is the construction always possible?

Chapter 2

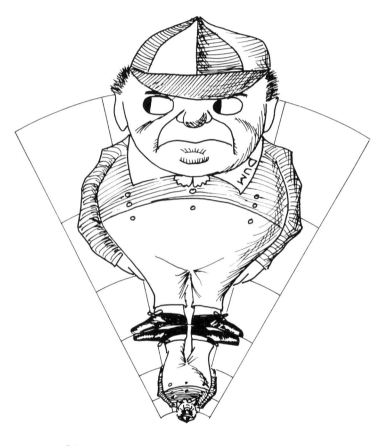

Reflections on Inversion

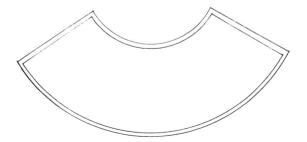

THEY are both studying maps," thought Alice, looking at Lewis Carroll and Whatif as she entered the room. "I would have thought that they already *know* where they want to take me."

The two men were engaged in discussion and did not seem to take any notice of Alice.

> "We'll use the Mercator
> When near the Equator
> But, what if we get near a pole?"

> "*This* chart is more apt
> The way it is mapped
> To play as expected, its role."

Alice went nearer to see the maps they were holding. "But this is where you took your families of circles yesterday," she exclaimed.

The maps looked indeed like the nets of lines she had met the previous day. Moreover, when she went nearer, she noticed more detail: latitudes and longitudes were marked, and in the place where A had stood before there was now a sign S.

Lewis Carroll explained kindly, "S stands for South Pole. This is a *stereographic* map."

He took the globe off from its frame and proceeded with the explanation. "You make the map by standing the globe on your sheet of paper, with S at the point of contact. Next you join N, the North Pole, with points on the globe and produce these lines to cut the sheet of paper. The map of point A on the globe is A' on the sheet, the point where the line NA meets the sheet and so on."

"You cannot have a big enough sheet for mapping all the points of the Northern Hemisphere in that way," objected Alice.

"You are right," said Whatif, "but it serves very well for mapping Antarctica."

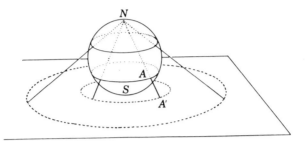

Alice looked at the two maps. The Mercator map with its straight perpendicular lines for latitudes and longitudes certainly looked more *natural* to her. She certainly had no intention of making a discovery trip to the South Pole. Miss Prim had even frowned when Alice once mentioned that she would like to travel to Paris.

"I *understand* why the latitudes on a stereographic map should be circles and the longitudes should be lines intersecting at the South Pole," she mused, "yet the picture does not come *alive* ... ."

Just then both maps suddenly came alive in an odd way.

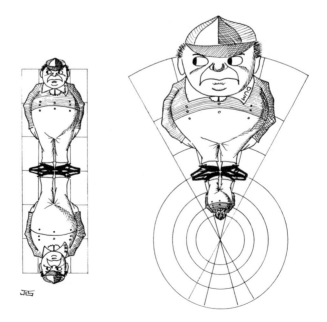

"Why, it's Tweedledee and Tweedledum!" exclaimed Alice, discovering long lost friends. They were unmistakably the fat twins, appearing on *both* maps at once. This time, however, the twins were separated. Tweedledee was on the Southern Hemisphere and Tweedledum in the North, placed quite *symmetrically* with respect to the Equator.

"You are always doing things upside down," shouted Tweedledum to Tweedledee. "Why are you standing on your head?"

"Contrariwise," shouted Tweedledee, "it is you who is standing on his head!"

The voices came from the Mercator map. Alice tried to mediate the quarrel.

"If you come to think, you are both right ... I mean, you are both wrong in saying that the other one is standing on his head. I mean, you are both right in thinking that you stand the right way. I mean ... ."

She was getting a little confused. She added hastily "Just think, reflect! You are just the reflections of each other."

"So we are, so we are," the twins shouted. This time, however, the voices came from the stereographic map.

"I did not mean it on *that* map. Why, Tweedlewee is much smaller than Twee-dledum. (Did I say Tweedlewee?)"

At this both twins produced tapes from their pockets and each measured his own girth. "Contrariwise," they both shouted at once, "I am exactly fifty inches."

Lewis Carroll took Alice aside. "They are *really* some kind of *reflections* of each other. Come Alice, it is time to do a little geometry. On our diagram, Q and Q' are two points on the same longitudinal circle and symmetrical about the Equator. P is the stereographic map of Q and P' the map of Q'. S, signifying the South Pole, is the centre of the stereographic map. E is a point on the Equator and T is its stereographic image."

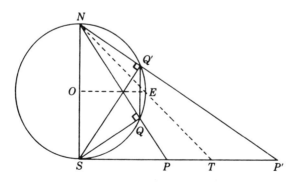

Alice looked at the diagram and exclaimed, "The map of E has gone pretty far from the centre. Why, the distance of E from O represents the radius of the Earth, but T lies *twice* as far from S."

"Really?"

"The midpoint theorem" Alice quoted hastily "in the triangle NST, O is the midpoint of NS, OE is parallel to ST, so ..."

"$ST = 2OE$" completed Whatif. "She seems to know her midpoint theorem."

"I also know all about the angles on the circle, and about similar triangles!"

Since Dr. Whatif had missed her performance on this subject the previous day, she had to show him now.

"In $\triangle NQS$ and $\triangle NSP$ the angle at N is common, and $\angle NQS$ is an angle in the semi-circle, so it is a right angle, same as $\angle NSP$, so

$$\triangle NQS \quad \text{and} \quad \triangle NSP \quad \text{are similar.}$$

"Good," said Lewis Carroll. "How about $\triangle NQ'S$?"

"$\triangle NQ'S$?" She drew a breath. "$\triangle NQ'S$ and $\triangle NQS$ are congruent; mirror-images in OE," she completed the argument.

"Can you see any triangle similar to $\triangle NQ'S$?"

"$\triangle NSP'$ is similar to $\triangle NQ'S$," she rushed on now, "two angles equal, just as in $\triangle NQS$ and $\triangle NSP$, and," she continued breathlessly, "it follows that $\triangle NSP$ is similar to $\triangle NP'S$, so

$$SP/NS = NS/SP'.$$

Now Lewis Carroll took over, "since $ST = NS$, the length representing the diameter of the earth, we can now write down

$$SP/ST = ST/SP'$$

or

$$\boxed{ST^2 = SP \cdot SP'}.$$

"Note that T is the stereographic map of E, a point on the equator," added Whatif.

"That is a very important result," said Lewis Carroll.

"Draw the circle **K** about S with radius $ST = k$. In our situation this circle is the map of the Equator. We call *P and P' points inverse with respect to the circle* **K**."

"So it turns out," continued Lewis Carroll, "that the stereographic images of the northern latitudinal circles are inverses of the images of the corresponding southern latitudinal circles."

"So Tweedledee and Tweedledum are actually the *inverse images* of each other," answered Alice. "You said that they were the *reflections* of each other," she added reproachfully.

"In a sense more general than you are used to think of reflection." This was Dr. Whatif joining the conversation.

"When you speak about the reflection of A, you think of the mirror image of A about a straight line ℓ. The line AA' is perpendicular to the reflecting line (or mirror) and the distances of A and A' from ℓ are the same. If P' is the inverse of P with respect to the circle $\mathbf{K}$, we say that P' *is the reflection of* P *in the circle* $\mathbf{K}$. The line PP' goes through S, the centre of $\mathbf{K}$, so PP' is still perpendicular (orthogonal) to $\mathbf{K}$, since the radius is perpendicular to the tangent at its endpoint."

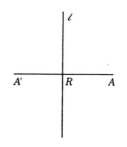

"But the distances TP and TP' are not equal!" objected Alice.

"No, they are not. Yet you will be able to convince yourself that if P and P' are *close* to the circumference of the circle, then PT and $P'T$ are *nearly equal*. You will also see that if we make the radius of $\mathbf{K}$ large, then the circle will approach a straight line and the inverse of point P in $\mathbf{K}$ will approach the reflection of P in the straight line. It will all be in your problem package. At a later stage you will see more clearly the close relationship between reflection and inversion. Now I ask you," he looked sternly at Alice, "have you done the last lot of problems?"

"I tried most of them," said Alice honestly, looking a little worried. "I think, I solved quite a few." She was still a little frightened of Dr. Whatif.

"Well, then Problem 9 should tell you how to find the inverse of a point P with respect to a circle $\mathbf{K}$. You can construct it using your compass and ruler. Doing the problem, you have seen that if you draw through point P circles orthogonal to circle $\mathbf{K}$, then they all meet in a point P' which is the *inverse* of P with respect to $\mathbf{K}$.

The relation $OP \cdot OP' = r^2$ means just that."

"So to construct P' it is sufficient to find *one* circle orthogonal to $\mathbf{K}$ through P and see where it cuts OP," said Alice.

"That is one way," said Lewis Carroll, "and apart from using the stereographic projection method, there are still other ways. You can do some as your homework."

"We would like to justify why we call an inversion a generalized reflection," added Whatif. "Let us try to get the inverse of a circle. I will however use one restriction for the time being. We will try to *invert about* $\mathbf{K}$ *a circle which does not pass through the centre of* $\mathbf{K}$."

Dr. Whatif continued, "Looking at the figure, suppose that $\mathbf{C}$ is the circle to be inverted and a line ℓ from the centre S of $\mathbf{K}$ intersects $\mathbf{C}$ in P and Q. Then the

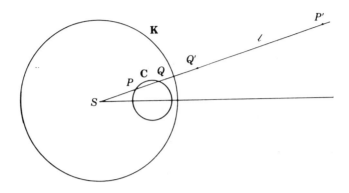

inverses of P and Q will be P' and Q', points on the line ℓ such that

$$SP \cdot SP' = k^2 \qquad \text{and} \qquad SQ \cdot SQ' = k^2 \qquad (k \text{ is the radius of } \mathbf{K}).$$

We have also the *power* of S with respect to $\mathbf{C}$:

$$S(\mathbf{C}) = SP \cdot SQ.$$

So dividing each of the two inversion equations by the last equation we obtain

$$\frac{SP'}{SQ} = \frac{SQ'}{SP} = \frac{k^2}{S(\mathbf{C})}.\text{''}$$

Alice forgot her manners and interrupted, "Why did you make a restriction on the circle $\mathbf{C}$?"

Lewis Carroll looked somewhat shocked at Alice. "Did you forget ..."

"Sorry for forgetting my manners."

"Nobody speaks about manners! Did you forget what $S(\mathbf{C})$ is if S is on the circle?"

"Why, it is zero!"

"Not even in the twentieth century do we go around dividing by zero," added Dr. Whatif scathingly, "let alone in Victorian times! So the case of S being on $\mathbf{C}$ will need special attention. Looking again at the equations we just got, you have that for *any* line ℓ cutting the circle in P and Q

$$\frac{SP'}{SQ} = c = \frac{SQ'}{SP}$$

where c stands for constant, since the number

$$c = \frac{k^2}{S(\mathbf{C})}$$

only depends on the circle of inversion $\mathbf{K}$ and the circle $\mathbf{C}$ to be inverted and *not* on the position of line ℓ. Say, for example, that $c = 4$. This means that P', which is the inverse of P, is 4 times as far from S as Q, while Q', the inverse of Q, is 4 times as far from S as P. The points, P, Q, P', Q' are on the same line through S. P and Q are both points of our circle $\mathbf{C}$, which is to be inverted. Can you guess now what will be the inverse image of the circle $\mathbf{C}$?"

After a little thinking Alice exclaimed, "Why, it is a 'blow up' of $\mathbf{C}$, everything will be 4 times enlarged! Only there is something peculiar."

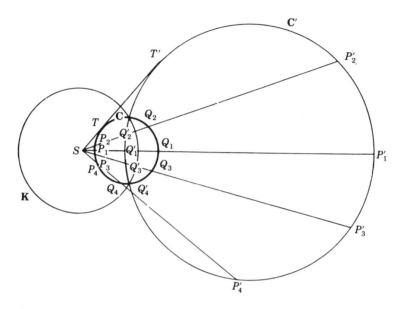

"Peculiar?"

"In the enlargement Q' corresponds to P and P' corresponds to Q."

"A good assessment of the situation," said Lewis Carroll, eager now to show off Alice to Whatif, especially after the previous blunder. "Of course, you must also consider the general case: $c = \dfrac{k^2}{S(\mathbf{C})}$, the factor of enlargement may take any value, and the circle $\mathbf{C}$ to be inverted about $\mathbf{K}$ could have different positions. It could be outside $\mathbf{K}$ or intersecting $\mathbf{K}$."

"If $\mathbf{C}$ is outside, $S(\mathbf{C})$ must be greater than k^2, so c is less than 1, and the inverse circle is inside $\mathbf{K}$ and smaller than $\mathbf{C}$," answered Alice.

"Sure. What if $\mathbf{C}$ intersects $\mathbf{K}$ at the points I and J?" asked Dr. Whatif, sketching a new figure. Alice was fast to observe, "Points I and J remain unchanged by the inversion."

"Are you sure?"

"By the inversion rule $SI \cdot SI' = k^2$. Since I is on $\mathbf{K}$, $SI = k$, so SI' must also be equal to k."

Lewis Carroll nodded. "Very good. A very important rule in inverting: *Any point on the circle of inversion remains unchanged.*"

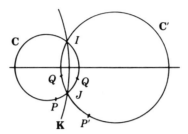

"This is a property," added Dr. Whatif, "where the circle **K** acts very similarly to a reflecting mirror. Points *on* the reflecting surface remain unchanged. One other look at the diagram brings home another property, the one you just observed: points marked with P of the circle **C** were *inside* circle **K**, and their inverses will be *outside* **K**. Points marked with Q on **C** are *outside* **K**, and *their* inverses on **C'** are *inside*. So again, **K** behaves a little like a mirror; it turns around the circle **C** (apart from making it larger or smaller according to the value of $c = \dfrac{k^2}{S(\mathbf{C})}.$)"

"Is it possible that an inversion does not enlarge the circle **C**?"

"Work this out for yourself!" urged Lewis Carroll.

"If $\dfrac{k^2}{S(\mathbf{C})} = \pm 1$," she exclaimed triumphantly, "*the circle does not change by inversion if it is orthogonal to the circle of inversion or if it coincides with the circle of inversion.*"

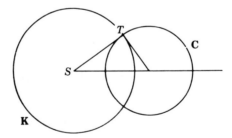

"Excellent," praised Whatif, "and note that the points on the internal arc have their inverses on the external arc and vice versa."

"This looks a little like a mirror image," said Alice. "But I would still like to know what happens if the circle **C** goes through the centre S," she mused.

"Try to think what happens when you try to invert S," helped Lewis Carroll.

"I cannot invert S," said Alice, "because the distance of S from S is zero, and I cannot divide k^2 by zero to get the distance of the inverse point!" She thought a little longer. "I cannot get it by the stereographic way either! S is the South Pole, its own image. The symmetric point on the Northern Hemisphere is N. N cannot have a stereographic image! Even points near N have *very far* images. Would the image of N be at infinity?" She looked really worried.

Lewis Carroll said soothingly, "You are finally getting the right idea, Alice. I think you even see now what Tweedledee and Tweedledum are up to. Just look at them again."

Alice looked as the fat twins moved about and it all came to her.

IN VERSE

Tweedledee and Tweedledum,
They are twin brothers stout.
One lives within circle **K**,
The other one without.

Yet ties of kinship, firm and strong,
Join brother to brother.
For each one of this couple is
The inverse of the other.

When Tweedledee moves inside,
Tweedledum moves out.
Tweedledee gets smaller,
Tweedledum more stout.

Beware the centre when you move,
Do not touch it Tweedledee.
Do not hurt your brother, Tweedle-
Dumb struck to infinity.

"Well," chuckled Dr. Whatif, "what if Tweedledee hits the centre with his nose? Tweedledum must take it with a *straight face.*"

"How can he?" asked Alice.

"He must. When a circle goes through infinity, it must go straight!"

Alice looked more and more bewildered and turned to Lewis Carroll for comfort.

"We must straighten you out first," said Lewis Carroll. "You just need some ordinary geometry to see that the inverse of a circle which goes through the centre of inversion is a straight line. Just look at the diagram." He drew **C** through the centre S. "Find first the diameter of **C** going through S. Let L be the other extremity and L' its inverse. Now pick a point P on **C** and find its inverse P'."

Alice gathered her thoughts, "$SL \cdot SL' = SP \cdot SP' = k^2$. I can see that much."

"Write it in the form $SL/SP' = SP/SL'$," helped Lewis Carroll. "Can you see any similar triangles?"

"Yes," triumphed Alice, "$\triangle SPL$ and $\triangle SL'P'$ have an angle common at S and the sides including the angles are proportional. So the other corresponding angles are equal! $\angle P'L'S = \angle LPS$. But $\angle LPS$ is in a semicircle, so it is a right angle.

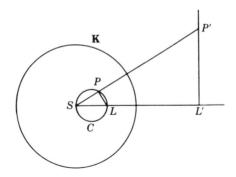

Therefore $\angle P'L'S$ is also a right angle."

"So?"

"So P' must be on the line perpendicular to SL' at L'. This line must be the inverse of **C**."

"Now you see it all. You can also see that when P runs around the circle **C** and gets very close to S, its image will be very far indeed on that perpendicular line."

Alice felt much happier, particularly because suddenly another idea came to her and she now understood her bewildering daydream which first sent her to Lewis Carroll's study. She exclaimed, "I see it now. We can also think of the circle **C** as being the inverse of the line $L'P'$. *The inverse of a straight line is a circle through the centre of inversion!*"

"If the *line* does not go through the centre," corrected Dr. Whatif.

Her odd dream suddenly made *sense* to Alice. "When I was standing straight in front of the *inverting mirror*, my image was an arc of a circle going through the centre of the inversion. As I moved away from the mirror, the inverted circle became smaller and smaller. I think, I now know how to invert *anything*."

"All right," smiled Dr. Whatif, "What if I ask you to invert these squares with respect to the circle. Much easier than squaring the circle."

Alice did not find the task too difficult and readily inverted both squares, the one touching the circle and the other one inscribed in the circle. "The inverses look pretty," said Alice, surveying with satisfaction her handiwork. "In fact, the inverses look prettier than the originals. I do not find enough *resemblance* in the inverses to call them *reflections* of the original squares."

"There is *some* important *resemblance*," remarked Lewis Carroll. "Look at the angles between the arcs NA' and $A'K$ on the first figure and between $N'A$ and AK' on the second one. Note that they are *right angles*, so they can be taken as reflections of the right angles $\angle NAK$."

"You may as well note that this is quite general," added Dr. Whatif. "*The angle between two lines remains unchanged by inversion*, though the sense of the rotation becomes opposite, just like on reflection in a straight mirror. You will do this in

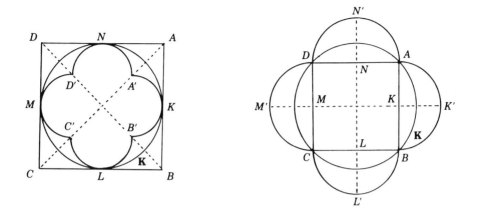

your exercise package. I give you the figure now and you can complete the proof."

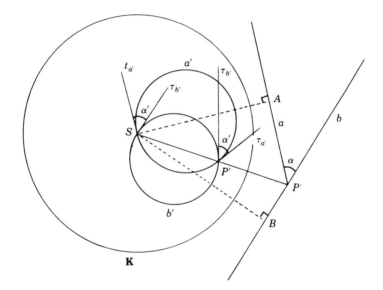

Looking at the figure Alice observed, "If the angles remain the same after inversion, that means that the inverted figure *looks* similar to its original as long as we do not go too far from the intersecting corners."

"You are getting the idea," said Lewis Carroll, "you are just about ready for your new exercise package."

"I wonder," mused Alice, "what happens if we invert two orthogonal families of circles? We must get another pair of orthogonal families, since the angles do not change."

"You've raised an interesting question. What will happen in particular if the centre of inversion is B, one of the intersection points of the elliptic family?" asked Dr. Whatif.

"All of the inverses of the circles in the elliptic family will become straight lines, since they all pass through the centre of inversion," said Alice.

"Very good," said Lewis Carroll. "Can you say anything about those lines?"

Alice thought for a minute. "All the circles went through A, so all their inverses must go through A', the inverse of A."

Lewis Carroll added, "If you wish, you can even arrange that A' should be the same as A. Simply let BA be the radius of the circle of inversion."

"What happens to the hyperbolic family?" asked Whatif.

"Think, Alice," helped Lewis Carroll, "the angles of intersection do not change under inversion!"

Alice saw it all perfectly well and resented the help a little.

"The only way the hyperbolic family can become orthogonal to the family of straight lines going through A is to become a family of concentric circles with centre A," said Alice very decisively, and she quickly drew the following picture.

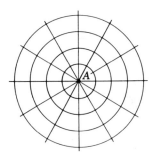

After drawing the picture, the two families looked very familiar indeed.

"But this is exactly the same situation we had previously when you ran away from point B to infinity, Dr. Whatif!" she exclaimed.

"Quite," said Whatif. "What if I come up with the following proposition: why not invite Infinity as *one* new point to join your Euclidean plane? It would make things very much easier!"

"Easier?" asked Alice.

"Sure. In your theory of inversion you have so many things to remember

the inverse of a circle is a circle, provided that it does not go through the centre of inversion;

the inverse of a straight line not through the centre, is a circle through the centre of inversion, its diameter perpendicular to the line;

the inverse of a circle through the centre is a line perpendicular to the diameter;

the inverse of a line through the centre is the line itself;

the centre of inversion has no inverse;
the North Pole has no stereographic projection.

Now, if you invite Infinity, then straight lines simply become circles going through it. Infinity serves well to become the inverse of the centre, or the image of the North Pole, or a meeting point of all lines you call straight in Euclidean geometry. You can say then that the inverse of every circle is a circle."

Alice looked convinced. "It is a good idea to invite Infinity. I always wanted to have a closer acquaintance with it, but my teachers were not very helpful, not even you, Uncle," she said reprovingly to Lewis Carroll. "I used to lie in bed, trying to figure out things about Infinity. One day . . .," her voice trailed off in a dreamy way, "one day, the Dodo-bird appeared suddenly and spoke to me, 'The way things are these days, I am getting out of practice with flying. I must take off again. Come and keep me company.' 'But I have never tried flying before,' I protested. 'As far as I *remember*,' said the Dodo, 'there is nothing to it. I go up and you just follow me.' With this he took off and I followed him. It *was* easy."

"We flew and we flew and we flew. Then I suddenly knew." Here Alice began to sing and because the tune was the familiar "My Bonny Lies over the Ocean," Lewis Carroll and Whatif soon joined her in the chorus.

I think we are over the ocean,
I think we are over the sea.
And somehow I have the odd notion
That we go to Infinity.

Going, going, here we all go to Infinity,
Going, going, going to Infinity.

We're passing so fast all the milestones,
I wish that I could only see.
I should hate so missing the *last* one,
The one that's marked: Infinity.

Going, going, here we all go to Infinity,
Going, going, going to Infinity.

The lonely and lovelorn lion *A*,
He wishes so to meet lion *B*.
Alas, they were *parallel* lions;
They met in Infinity.

Going, going, here we all go to Infinity,
Going, going, going to Infinity.

I worked and I worked at that problem,
I searched what the answer could be.
A voice rang out laughing and jeering:
"Find me in Infinity."

Going, going, here we all go to Infinity,
Going, going, going to Infinity.

Each point in the plane was *inverted*,
The centre cried, "what about me?"
Now, find for the centre an inverse;
Let him have Infinity.

Going, going, here we all go to Infinity,
Going, going, going to Infinity.

"Don't go yet Alice," cried Lewis Carroll. "Don't forget to take your package of problems!"

𝔓roblems and Exercises 2

1. Let P be a point inside the circle **K** and distinct from its centre S. Let A be a point on the circumference of **K** such that AP is perpendicular to SP.

Let the tangent to the circle at A intersect SP produced at P'. Prove that P and P' are inverse with respect to the circle **K**.

Hence find a construction for the inverse of a point with respect to a circle, if the point is

a. inside the circle;

b. outside the circle.

2.

a. Two circles C_1 and C_2 *touch* each other at the centre S of the circle **K**. Find the inverses of C_1 and C_2 with respect to **K**.

b. Find the inverses of two parallel lines with respect to the circle **K** if neither of the lines goes through the centre S.

3. Show that the inversion leaves the angles between two lines (circles) unaltered. What can you say about the inverses of two tangential circles?

Hint. Use figure given in the text on page 39.

4. Given the circle **K** with centre S, find inverses to

1. a line, tangent to **K**;

2. a *segment* of the tangent line;

3. a chord of the circle;

4. a diameter of the circle.

5. The circle **K** is inscribed in a polygon of n sides so that it touches the sides of the polygon. Determine the inverse of the polygon with respect to **K** and show that the length of the perimeter of the inverse of the given polygon is equal to the circumference of **K**, irrespective of the value of n.

6. P is a point inside the circle **K**, of centre S and radius k. Let P' be the inverse of P with respect to **K**. Let SP intersect the circle **K** at A. Denote the distances PA, AP' by x, y respectively.

Let $p = x/k$ (i.e., the ratio of the distance AP and the radius). Show that if p tends to 0, then y/x tends to 1. (This means that a point and its inverse become very nearly equidistant from the circumference of the inverting circle either if the points are close to the circumference, or if the radius of the circle is increased to make the circle approximate a straight line.)

7. The circle **C** is inverted with respect to circle **K**. Under what condition is the centre of the inverse of **C** the same as the inverse of the centre of **C**?

8. A circle C_1 is completely inside the circle C_2 but the centres of C_1 and C_2 are different. Is there an inversion such that the inverses of C_1 and C_2 are *concentric*?

Hint. Regard C_1 and C_2 as members of a hyperbolic family.

9. (Steiner's porism). Two circles C_1 and C_2 are such that C_2 is completely inside C_1. Show that *if it is at all possible* to place *a chain of tangential circles between the two circles* so that each of the circles touches both of its neighbours, and both given circles are as shown on the diagram, then it is possible to do this in an infinite number of ways. The number of circles in the chain, however, is the same for all solutions.

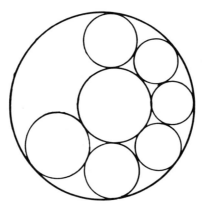

10. K is a circle of radius k and centre S and ℓ is a line not going through S. A and B are two points on a line through S and perpendicular to ℓ, such that A and B are equidistant from ℓ. Let A', B', ℓ' be the inverses of A, B, ℓ, respectively. What can you say about A', B', and ℓ'?

Dr. Whatif's

Euclidean

Geometry

EWIS Carroll opened the door and let Alice ahead into a large, splendid hall. It looked rather like a ballroom with its highly polished floor. Dr. Whatif was standing near the centre of the room, beckoning them to come forward. He was pointing with his elegant walking stick to the pattern showing on the floor.

Alice looked at it closely. The floor pattern consisted of beautiful, shining circles, some large, some small, all interlocking at a point, clearly marked O. They seemed to be inviting for a waltz. "What a pity!" thought Alice to herself. "Here I am in this beautiful ballroom, with nobody to dance with. I could not think of Dr. Whatif as an exciting partner, nor of Uncle L.C.," she mused somewhat ungratefully. Just then Dr. Whatif interrupted the train of her frivolous thoughts. He pointed to the point O, just at his feet, where all the circles met.

"This is going to vanish," he said. He pulled out a little spray-can, sprayed something from it and lo and behold, the point O was obliterated. By now, Alice was used to Dr. Whatif's strange doings. This time, however, he did not fly away to infinity and the shining circles on the floor remained untouched.

For a moment Alice had some doubts.

"You have wiped out the letter O, *denoting* the meeting point of all these circles, but can you really *remove* a point?"

"The point is not there," said Dr. Whatif quite emphatically. "Of course, this time, even your *extraordinary* good eyesight cannot decide whether it is there or not, so we put down a reminder."

With these words he pulled out a little flag and stuck it down at the meeting of the shiny circles.

"We are going to have a new geometry here," he announced. "We will have points and lines, just as befits a geometry. Our points will be all the points still left here" (Alice glanced at the little monument mourning the lost *O*) "and *our lines will be all these circles emerging from under the flag.*"

"They do not *look* like lines to me!" said Alice.

"Not even this one?" asked Dr. Whatif, pointing to one of the shiny tracks, which looked *straight* even to Alice's extraordinarily good eyes. She soon discovered many other lines, all pointing radially outwards, from the meeting nonpoint.

"I do not know, if these are *really* straight lines or just arcs of very large circles."

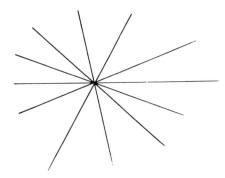

"Ah, but that does not make any difference since yesterday when we invited Infinity to join our plane," said Dr. Whatif, "what you *call* straight lines, are simply circles going through the point 'infinity'."

"Anyway," continued Dr. Whatif, "in our generalised geometries we do not really care very much, what things *look* like. Our eyesight after all is very limited."

Lewis Carroll wanted to object, but Dr. Wahtif continued relentlessly, "We set up our axioms and investigate the consequences."

"But ..."

"All right, I know that you have very firmly ingrained ideas about *your* geometry. I will make a concession to you and will call our structure a *model of geometry*, and directly you will see that things are much the same in our model as in your Euclidean plane geometry."

"In Euclidean geometry," said Lewis Carroll, "there is *exactly one line through two distinct points.*"

Dr. Whatif beamed, "And I ask, how many circles can you draw through three given points?"

Here Alice answered eagerly, "I can get exactly *one* circle and I can show you how I construct it."

Somehow, even before she finished the sentence, the sheet with the completed sketch appeared in her hand, just as ready as Pallas Athene when she leapt in her full armour straight out of the head of Zeus.

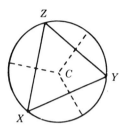

"Clearly," she said, "the perpendicular bisectors of the sides XY and YZ can only meet in *one* point, which will be C, the centre of the circle."

"Are you sure," asked Lewis Carroll, "that those perpendicular bisectors intersect at *all*?"

"They cannot be parallel!" exclaimed Alice.

"Unless ...," said Lewis Carroll.

"Unless," continued Alice quickly, getting the idea, "X, Y, Z are on the same straight line."

She glanced at Whatif and remembered that mysterious point at Infinity. "In that case, however, we already have a line which is a circle through Infinity."

"Very good!" said Dr. Whatif in an appreciative tone which, from him, was quite new. "You can now see the rules."

"They seem to me like the rules of a funny game. You invited Infinity, but expelled the point O."

"That is just the point I was going to make," said Dr. Whatif.

"I thought it was the point you just *unmade*," interjected Lewis Carroll.

"The point I am making," said Whatif, ignoring Lewis Carroll's pun (What else can you do about puns?), "is that through each of *our two points*, and remember, there is *no O*, there goes just *one* of *our lines*, which are, ..." he looked inquiringly at Alice.

"Circles emerging from under the flag" said Alice.

"I admit, that is true," said Lewis Carroll, "but what about the intersection of lines? In Euclidean geometry two lines intersect in at most one point."

"Well," said Alice, "two circles intersect in at most *two* points!"

"But our *lines* are very special circles as you have observed before."

"True, they are a little ... mutilated," said Alice, "so each pair of circles has only one intersection point left."

"They may not have *any* intersection points left," corrected Lewis Carroll. "How about the two fellows on the left? Or the two on the right?"

"Good!" said Whatif, "so we have also some *parallel* lines."

"Oh, I would not be too hasty in calling two lines parallel, just because they do not intersect," said Lewis Carroll. "In Euclidean geometry they are rather special. Through a point P not on the line ℓ you can draw exactly *one* line parallel to ℓ. We should check that this is the case here."

"That will be a little problem for Alice to do later," said Whatif. "She will have to prove that there is exactly one circle which touches a given circle at a given point and passes through another given point not on the circle."

Lewis Carroll seemed to be quite satisfied. He added, "We can draw the same conclusion as in Euclidean geometry: *If the line a is parallel to the line b, and the line b is parallel to the line c, and a and c are distinct, then a and c are parallel.*"

"We can conclude this?" asked Alice. "How?"

"Think, girl, think! Maybe you can answer the question yourself."

"Well, if a and c are distinct and not parallel, then they meet in a point P. But then we have through the point P, *two* lines parallel to the same line b. This cannot happen, since there is exactly *one* line through P, parallel to b."

"Maybe, after all, this girl *can* think," said Whatif.

Alice looked indignant.

Dr. Whatif produced the following sketch and asked, "Now, how about this situation?"

Alice noticed a suspicious twinkle in his eyes, so she ordered all her braincells on alert.

"Here you have the lines a, b, c as denoted. Now a is parallel to b, b is parallel to c."

Alice raised her eyebrows.

"Well, you would not say that b and c meet since there is *no* point O. So now a is not parallel to c, although a, b and b, c are two 'parallel' pairs," teased Whatif.

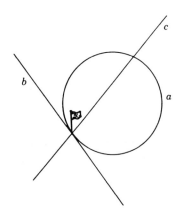

Alice broke out in song:

> "Going, going, here we all go to Infinity,
> Going, going, going to Infinity.
>
> There's no point called O right nearby,
> A place where line b could meet c,
> But do not take *no* for an answer,
> They'll meet in Infinity."

"So b and c are not parallel!"

"That song puts us in a happy mood," said Whatif. "We have established that in our geometry:

(1) two distinct points determine exactly one line, hence two distinct lines intersect in at most one point;

(2) through a point P not on a line ℓ there is exactly one line parallel to the line ℓ."

"Oh," said Alice, "there is so much more in geometry than this! There are theorems about perpendicular lines, there are ..."

"Alice," reprimanded Lewis Carroll, "you did not forget your families of orthogonal circles! They made such a strong impression on you."

"The Montagues and the Capulets?"

"It would be more to the point," said Whatif in a sneering tone, "for you to show that in *our* geometry there exists exactly one line perpendicular to a given line at any point of the line; also how to construct a perpendicular from a point to a line."

"We can leave this for her problem package."

Alice was thoughtful. "Of course, here the lines are those special circles through nonpoint O, but ... if you call those circles lines, are there any *circles* in this geometry, I mean, are there any things you can *call* circles and what do they look like?"

"Now, that is a well justified question," said Lewis Carroll.

"And I am sure that the young lady will get the answer," said Dr. Whatif "with just a little *reflection*." He turned to Alice and said, "Suppose that in your old-fashioned Euclidean geometry you have a family of lines through a point C. We call them sometimes a pencil of lines. Suppose that P is a point distinct from C. What is a good way of finding a circle about C through the point P? REFLECT!"

Alice caught the idea. "This has not occurred to me before, but I see that I could get many points of the circle by reflecting P about those lines through C, or by reflecting repeatedly the images of P. All those images lie on the circle."

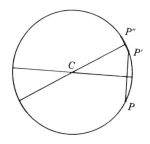

"And all the points on the circle can be regarded as the reflected images of P about some of the lines of the pencil."

"Moreover," observed Alice, "the circle so obtained will be *orthogonal* to the given pencil of lines."

"Good, good!" Dr. Whatif was pleased with her. He continued, "Now, what if, ... what if we come back to *our* geometry here and consider a family of *our* lines through a point S. Can you do anything about reflections now?"

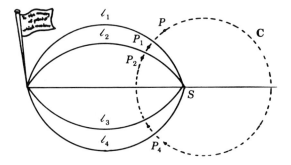

"Reflections? In this case the reflections are inversions ..."

Both men nodded. She took courage, "Why, if P_1, P_2, P_3, P_4 are the reflections, (inversions) of P about $\ell_1, \ell_2, \ell_3, \ell_4, \ldots$, they all lie on a circle, a real true Euclidean circle $\mathbf{C}$, which is moreover orthogonal to all those lines! But ...."

"Good, good," said Dr. Whatif. "You will prove it all in detail later."

"But S is not the centre!" cried out Alice and Lewis Carroll together.

"It is not the *Euclidean* centre," admitted Whatif. "The time has come to introduce *new scales* for our geometry."

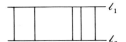

"For that matter," added Lewis Carroll, "let us return to the lines you call 'parallel' in this geometry. In Euclidean geometry the perpendicular distance between two parallel lines remains the same wherever you take the measurements. You would *not* say that that holds in this situation!"

"I certainly would!" countered Whatif. "We must use methods of measurement to fit our geometry. In Euclidean geometry a rod and its reflected image have the same length. Here too we will use the principle of reflection to compare lengths. We can show that the line-segments A_1A_2 and B_1B_2 in your sketch are inverse to each other about some line of *our* geometry."

"Are $A_1A_2, B_1B_2, \ldots$ arcs of circles belonging to the family orthogonal to the family of ℓ_1 and ℓ_2?" asked Alice.

"After our merry day with orthogonal families of circles you should be quite sure

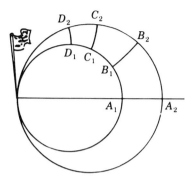

of this," said Lewis Carroll. "Here we have the limiting case at hand, when both families, the elliptic family, all members of which intersect in two points, and the hyperbolic family, of which no two circles intersect, each become families touching in two coincident points. To your collection of pairs of orthogonal families add the picture on the left."

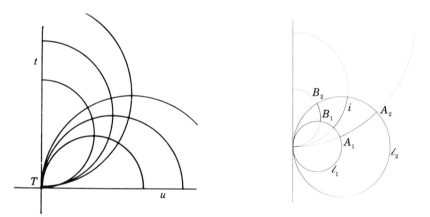

"It is pretty," said Alice.

"Look now at the picture on the right. If the members of the hyperbolic family move closer to touch their radical axis h at T, the members of the elliptic family are forced to fuse the two points common with *their* radical axis c into a single point."

"And it is the *same* point T," exclaimed Alice. "So the Montagues and Capulets meet after all."

"In the case we are discussing right now, the two families touch at the nonpoint O!" continued Dr. Whatif. "We can add another line i, which you would call a Euclidean circle, to the picture. I claim that whatever way you choose the segments A_1A_2 and B_1B_2, orthogonally to ℓ_1 and ℓ_2, we can find the line i so that B_1B_2 is the inverse of A_1A_2 about i. That is to say that B_1B_2 is the *reflection* of A_1A_2 about i, hence in our new scale, *equal in length* to A_1A_2." (See figure on the right above.)

Alice had suddenly become very tired. Those lines, still not looking at all like lines, were shining in her eyes. The task of measuring distances between them or along the lines became almost intolerable. It was as if a heavy weight pressed down on her head and pulled her eyelids. She sank down to the floor watching the circles and she put her hand to her forehead in deep concentration. Looking closely at the shining lines she suddenly discovered more details. Shapes emerged along the lines: they seemed to be flowers; yes, they were *pressed* flowers. Her own favourites, specimens she had carefully collected and treasured. She also heard suddenly a voice singing, it was not the voice of Lewis Carroll or of Dr. Whatif:

> "Snip, snap, prune
> Morn to afternoon
> Sing a merry tune.
> Equality's the principle we always emphasise,
> The garden must be *uniform*, nothing unprecise.
> Nothing tangled, nothing sloppy,
> Every plant a faithful copy,
> Chop the tall, unruly poppy,
> Cut it down to size.
>
> Sing a merry tune
> Snip, snap, prune"

She recognised now the singer. He was a card, one of the gardeners of the Queen of Hearts. He was brandishing a pair of shears. However, to Alice's eyes the flowers did not look as uniform as the song claimed.

"Good afternoon," she addressed the gardener.

"At last you notice me," growled the card. "You have been sending here those pressed flowers for a long time. I have been tending them faithfully, but it is time that you joined me here."

To tell the truth, Alice had felt rather *flat*, so this exercise did not seem too hard for her.

"Come and check yourself that all those flowers are of the right height. Just come and measure them against yourself."

Alice wanted to protest. For one thing, she *knew* that the flowers were of different size. She also found the idea of measuring them against herself quite ridiculous. There was no "drink me" bottle anywhere in sight which could reduce her own height.

"Why do you hesitate? You could go through the looking glass before, couldn't you?"

"Well, it could not be really too difficult" she reflected. Yes, she *reflected* and found herself hugging a tall poppy, it was just her own height. She reflected again. "It seems that the gardener is doing well the task he set out for himself," she thought as she hugged the next poppy—and the next one and the next. She bounced and reflected, she reflected and bounced again, again and again. Then suddenly there were no more flowers. Still she went on bouncing and reflecting, on and on. "Why, I must have been reflecting for *hours*," she thought. "Surely by now I must have reached point *O*!"

"Shh, there is *no* point *O*," said a voice as if reading her thoughts.

Alice turned in the direction of the sound, it was Tweedledee! It was really Tweedledee. Next to him stood a strange box. Alice stared at the box, for she had never seen anything like it.

Tweedledee laughed "I must keep in contact with Tweedledum. This is a television receiver."

"Indeed," Alice gasped with surprise. On the screen she saw the image of Tweedledum.

Tweedledum was laughing, "Big deal! Dr. Whatif speaks about a new model of a Euclidean geometry. Can't you see that he is pulling your leg? He is presenting you with the *inverse* image of the Euclidean plane with its lines and points. No wonder everything works just *like* in Euclidean geometry! You and my brother Tweedledee will get to your nonpoint *O* at the same time *I get to Infinity*. Hahaha .... ."

"Hahaha," laughed Lewis Carroll, "this inverse image gives an amusing model of Euclidean geometry."

Alice had her eyes wide open. There was no sign of Tweedledee, or of the pressed flowers, or of the card gardener. The two men were involved in a discussion.

"Well, it is a neat little trick," admitted Whatif. "However, I think I have made my point. I have now shown that given some sets of objects which we call points and lines, though they do not *look* like the Euclidean lines we can apply the Euclidean axioms, or rather a set of equivalent axioms including the parallel axiom, to them and I can build up Euclidean geometry. Alice will be able to prove some theorems in detail."

"What about consistency?"

"All our deductions can be proved, since this *model* is really Euclidean. So if you

really believe that Euclidean geometry is consistent, and after all those centuries of testing we feel quite justified in believing that it is, then you must accept the validity of *our* geometry."

"It *looks* different, but as you say there is nothing really new in it!"

"Nothing really new in this geometry. But you and Alice are now ready to accept more general, new geometries, if we verify the consistency of the axioms with some other Euclidean model."

Alice was a little bewildered and tired too. She could not suppress a yawn.

"We should call it a day," said Lewis Carroll kindly.

"But do not miss your new problems after having your rest," said Dr. Whatif, handing to Alice the fresh problem package.

𝔓roblems and 𝔈xercises 3

In what follows, D.W.E.G. stands for "Dr. Whatif's Euclidean Geometry."

1. Let **L** be a line in D.W.E.G. and A a point on **L**. Show how to construct through A, a line p, perpendicular to **L**.

Interpretation. Let **L** be a circle going through O and A a point on **L** distinct from O. Construct a circle through A and O, orthogonal to **L**. Is the construction always possible?

2. Let **L** be a line in D.W.E.G. and A a point, not on **L**. Construct a line through A that is perpendicular to **L**.

Interpretation. Let **L** be a circle through O and A, a point not on the circle **L**. Construct a circle through the points O and A that is orthogonal to **L**. Is the construction always possible?

3. Show that the sum of the angles of any triangle in D.W.E.G. is $180°$.

Interpretation. A triangle in D.W.E.G. is a figure bounded by three circular arcs AB, BC, CA each belonging to a circle through O; A, B, C being distinct from O. The angles of the triangle are the angles between the tangents to the circles drawn at A, B, C, respectively.

4. Let $\mathcal{E}$ be a family of (Euclidean) circles intersecting at the points O and P, and let A be a point distinct from O and P. Show how to construct a circle **C** through A, orthogonal to all the circles of the family $\mathcal{E}$.

Hint. see Chapter 1, Problem 2.

5. Let P and A be distinct points in D.W.E.G. A *circle of D.W.E.G.* about the point P (as centre) and through the point A is defined as the locus of the reflections of A in the (D.W.E.G.) lines through P. Show that such a circle is also a circle in ordinary Euclidean geometry, though the point P is not the Euclidean centre.

Intepretation. A line through P in D.W.E.G. is a Euclidean circle **L** through O and P. The reflection of A in **L** is the inverse image of A in the Euclidean circle **L**.

    a. Let **C** be the circle defined in Problem 4, i.e., a circle through A, orthogonal to the family $\mathcal{E}$. Show that if **L** is any circle in $\mathcal{E}$ (i.e., any line of D.W.E.G. through P), then the inverse of A in **L** lies on **C**.

    b. Show that if A' is any point on **C**, then we can find a circle **L** of $\mathcal{E}$ (i.e., a line of D.W.E.G. through P), such that A' is the inverse of A in **L**.

From (a) and (b) it follows that the Euclidean circle **C** is a circle in D.W.E.G., *about the point* P, through the point A.

6. Show that if **L** is a line in D.W.E.G. and A a point not on **L**, then there exists exactly one line through A, parallel to **L**.

Interpretation. **L** is a circle through O, and A a point not on **L**. Find a circle through A, touching **L** at O. Is the construction always possible? Is the solution unique?

7. A *parabolic family* of circles is defined as a set of circles with a common point of contact. Let $\mathcal{P}_1$ be a parabolic family touching at the point T, with the common tangent line t. Let u be a line through T, perpendicular to t. Show that the family $\mathcal{P}_2$, consisting of circles touching the line u at the point T, is orthogonal to the family $\mathcal{P}_1$.

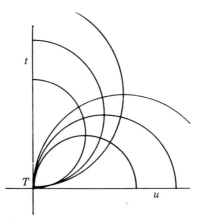

8. Let **L** and **M** be two parallel lines of D.W.E.G. A and B are points on **L**, C and D are points on **M** such that the line-segments AC and BD are perpendicular to **L**.

    a. Show that AC and BD are perpendicular to **M**.

    b. Show that the segment BD is the reflection of the segment AC about some line **N**.

Interpretation. **L** and **M** are two circles *touching* at O. A, B are points on **L**, C, D on **M** such that the circles through O, A, C and O, B, D respectively are orthogonal to **L**.

    a. Show that the circles OAC, OBD are also orthogonal to **M**.

    b. Find a circle **N** through O such that the arc BD is the inverse of the arc AC in circle **N**. (BD and AC are arcs of the circles OBD and OAC.)

Chapter 4

A Hyperbolic T-Party

H E is the MAD HATTER!" Alice thought suddenly, looking at Dr. Whatif. To be sure, he was not wearing a hat, but Alice was reminded of that mad tea party of so long ago, when she found herself seated between him and someone else. ("No, it could not be L.C., *he* would not look quite so sheepish.")

On a closer look however, the creature was not a sheep, rather a hare. "Of course, he must be the March Hare!" However, the similarity to that old tea party ended here. For one thing, she could not see the ends of the table at all. She ventured the question, "I presume this is a tea party, but I cannot see any cups."

"You do not need any cups," said Dr. Whatif. "There is some T for everybody who cares to come to this table."

Indeed, Alice saw now that there was a T at every setting, though the other side of the table got her somewhat worried. At the farther sides the positions of the T's looked a little precarious. "They may topple off any minute," she thought.

She turned to her other neighbour, who seemed to be the host. "Excuse me for asking. Where are the ends of this table?"

"Oh," answered the creature ("yes, he is definitely the March Hare," thought Alice). "You have very old fashioned ideas about tables. You can only think of rectangular table tops. Well, here we simply *do not believe in rectangles*. Besides, we are very thrifty here, we do not pay carpenters to finish the ends. We make both ends meet."

"Nonsense," said Alice. "How can you make *both* ends meet? Two *straight* lines can only meet in one point!"

"You have the point, young lady," said the Nonhatter, who really *was* Dr. Whatif,

61

"my friend has some hary ideas! In fact, the ends do not meet at all."

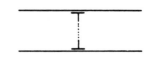

He joined Alice's T with the opposite one, using a fine straight thread. "This line is perpendicular to both edges, so they *must* be parallel."

"Sure," said Alice, all her Euclidean learning coming to life, "the two lines are parallel, if the transversal forms equal interior angles with the lines." Whatif nodded. Alice continued, "and if the two lines are parallel, then those alternating angles are equal."

"Not so fast."

"You do not mean this."

The voices came in unison from all around the table. Startled, Alice said, "but I mean what I say," and to be more convincing she pointed to the line joining

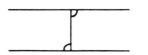

the two sides of the table. "If the two edges could meet on the right, then by symmetry they'd meet on the left and you all must agree that two lines cannot meet in two points!

To modern times from the times of the Greek
A line through two points must be unique."

Since Alice met no disapproval now, she continued, "So if two lines are parallel, those alternating angles must be equal, to say what I mean."

"It is the same, eh," sneered her neighbour (the Hatter?), "to say what you mean and to mean what you say?"

The Hare added, "You like what you eat and you eat what you like, all the same?"

Alice retorted, "Since I do not eat anything, I cannot *dislike* it. As to eating what I like ...." She pointed to the table which did not offer anything apart from those T's.

"Well," said Whatif, "your friend, the venerable L.C. is not here. Even he would object to your logic, though he has blinkers on his eyes and cannot see beyond Euclid and his unique parallel through a point."

"Where is Lewis Carroll?" asked Alice, dismayed by his absence.

"He is not big enough to be at our party."

"Not big enough?"

"You may have been too sleepy to remember the drink that all of us here have drunk to make us larger. This table is so much larger than the one Lewis Carroll and his Oxford friends are used to. In fact, this table is part of a *hyperbolic plane!*"

"Of what?"

Suddenly Whatif produced a huge protractor and drew a triangle on the table. "This is an *extraordinarily* precise protractor. Now use your extraordinary eyesight to measure its angles."

The triangle and protractor did not look very special to Alice. "I wonder if they were also soaked in that drink," she thought.

She measured:

$\alpha = 84°28'$

$\beta = 65°07'$

$\gamma = 28°05'$

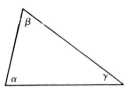

She checked again and having obtained the same, added for a further check. She cried out, "$\alpha + \beta + \gamma = 177°40'$! Something must be wrong."

Whatif winked mischievously and added another triangle.

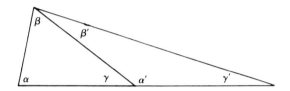

"Just for a check, measure α', β', and γ'"

" $\alpha' = 151°55'$ —just as I expected.

$\beta' = 14°16'$

$\gamma' = 13°20'$

The sum is now $179°31'$. Not as bad as before, but it still cannot be right!"

"Measure the angles of the big triangle," egged on Whatif.

Alice measured: "$\alpha = 84°28'$ —just as before,

$\beta + \beta' = 79°23'$ —checks with the rest,

$\gamma' = 13°20'$ —as before.

The angle sum is now $177°11'$. This is worse than before!"

"Perhaps the angle sum is *really* less than 180°. Besides, it is not hard to show that if $\alpha + \beta + \gamma$ and $\alpha' + \beta' + \gamma'$ are both less than 180° you cannot expect the sum of α, $\beta + \beta'$ and γ' to make 180°," argued Dr. Whatif.

"Oh, Dr. Whatif, what if, what if, . . ., what if the angle sum turns out to be *more* than 180°?"

"*That* cannot happen!" announced Dr. Whatif with an air of assurance which made Alice reluctant to contradict and yet she did not feel convinced. Whatif sensed her bewilderment and laughed. "You will come round to see this yourself doing your thinking with a clearer head. It will be all in your problem package. It will involve a pretty drawing like this, with many copies of the triangle $B_1 A_1 B_2$ which *claims* to have more than 180° for its angle sum. But I will leave this to you. For the time being take this theorem as an advance loan."

"Then how can the angle sum be less than 180°." Then she added, remembering

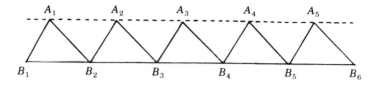

suddenly her school geometry, "It is easy to prove that it
must be 180°. You just draw a parallel through the vertex
and mark in the equal alternating angles. There you are!"

"Oh, there we are back to those parallel lines," answered
Whatif, "I admit that you can prove that if the alternating
angles are equal then those lines must be parallel, but *nobody* could prove the
converse. This is why Euclid put the converse (or what is equivalent to it), as his
famous fifth postulate for the Euclidean plane. But now we are in a . . . ."

"diabolic plane?" asked Alice.

The Hare shook his head violently, his whiskers trembled. "How rude to interrupt!"

"Wrong too," said Whatif, "the word is *hyperbolic*."

"Whatever it is . . . ."

"It may not appeal to your square Victorian thinking," teased Whatif.

"She is a square," cried the Hare, "we cannot have any squares here!"

"Not even rectangles," continued Whatif.

"This table is not a rectangle, because the Hare was too mean to finish it," said
Alice angrily.

"You could not even draw any rectangles on it," said Whatif.

Alice grabbed some drawing instruments which seemed to take the place of butter
and jam on that table and said, "but I can draw one." To begin with she drew the
figure

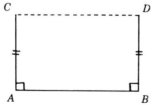

"Surely, I could make right angles at A and B. I could make AC congruent to
BD!"

Dr. Whatif nodded. "Now join C and D, grab that protractor and measure the
angles at C and D!" This was Dr. Whatif's command.

"Oh, they are both equal . . . ."

"Equal to what?" asked Whatif sneeringly.

"86°24'."

"I congratulate you again on your extraordinary eyesight!"

"And on the precision of the protractor, which is supplied by the house," added the Hare.

To be sure of the protractor, Alice checked the angles at A and B. They were precisely 90°.

"In this plane there are no rectangles," said Whatif.

"Maybe, if I tried long enough, I could find some. It is odd enough that this *one* is not a rectangle, but .... ."

"Out of the question! You have found here a triangle with an angle sum less than 180°."

"She found three," cried the Hare.

"But, perhaps I could find a good triangle." Alice was hopeful.

"What do you mean, one without an *angular defect?*"

"Yes."

"All right, let us *suppose* that you can find one. In that case I admit even that you can find three triangles without angular defect."

Alice brightened. "Yes, because I could chop my triangle into two triangles which have no defect. I can even make them right angled triangles."

"Are you sure?"

"Yes, taking your word for it that there is no triangle with angle sum more than 180°, I could not have any one of the smaller triangles—er—defective—because then they could not add up to a perfect triangle, I mean one with no defect."

"Quite right, quite right."

Alice continued eagerly, "But then I can form a rectangle by putting together one of my *good* right angled triangles with a congruent triangle, like this:"

"What is more," continued Dr. Whatif, "you can make many other rectangles, you can make a rectangle as large as you like, by putting together like bricks rectangles congruent to the one you have found!"

"So, I could be right."

"Ah, but now comes the fly in the ointment! You remember that you have also found a defective triangle!"

"So?"

"So, you could chop it up into two right angled triangles, at least one of them must be defective."

"I suppose so."

"Now fit your little defective triangle in that shaded corner of your big rectangle and cut your rectangle into triangles as shown. Remember, your little triangle is "defective" and the others cannot have their angle sum exceeding 180°. So do a bit of arithmetic to see that you have no hope for producing four right angles on that rectangle of yours!"

"So that one little defective triangle spoils it all!"

"It certainly does. You cannot have your big rectangle, and going back to the beginning of our story, you cannot have any rectangles, any right angled triangles without defect, or any other triangle without defect!"

"You mean that in a plane *all* triangles are without defect, or *none* of them are!" wondered Alice.

"Precisely!"

"But I still cannot see that I was wrong when I proved that the angle sum must be 180°."

"No, you were not wrong, provided that the parallel postulate holds. In the Euclidean plane we *assume* that there is only one line parallel through a point to another line. But where we are now ... ."

"Things are different?" guessed Alice.

"Take again that defective triangle that you have found, and place the angles α and γ at vertex B as shown. Then by the alternate angle theorem you have *two* parallel rays issuing from B instead of one!" explained Whatif.

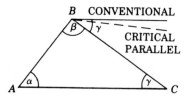

"What about the rays coming out of B *between* those two?"

At this moment something unexpected happened. For an instant the room was plunged into darkness. Then, simultaneously two bright fluorescent lines appeared, a red one joining Alice's T to the T on the opposite side of the table and a long blue one, all along the edge of the table, on one side of Alice. Alice gasped, fluorescent lights, of course, were quite new to her.

Then gradually the lights fanned out, until one half of the table was bathed in red light, the other half in blue light, with a thin silvery line dividing the two regions. Alice was fascinated. "Is this the answer to my question?"

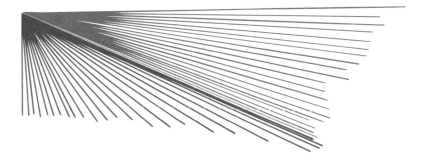

A thin voice came from the red region, "Why don't you ask us?"

Another tingling voice from the blue region, "or us?"

By now Alice was used to the unusual. She looked in the direction of the voices. She strained her extraordinary eyes to see the speaking slim creatures. "Are you eels, snakes, sprites?"

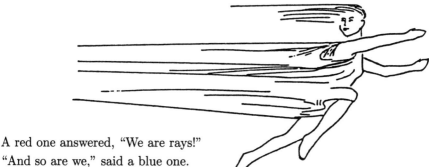

A red one answered, "We are rays!"

"And so are we," said a blue one.

"But you *look* different," said Alice.

"We are," said the red ray, "we belong to different clubs."

The blue one countered, "Never would we tolerate one of *them* between two of *us*."

"Nor we would have one of *them* between two of *us*," laughed a red ray.

"They!" the red ray continued scathingly, "they are the ones who could never reach the other side of the table."

"You mean, they are rays that are parallel to the other edge?" asked Alice.

"We all are," said the blue one, "and proud of it too!"

Alice was a little sceptical. "I can see why a parallel ray can be proud in the Euclidean plane, he is unique, unparalleled. But here? Perhaps I could talk to the silvery ray, "she thought.

"Do *you* belong to either of the two clubs?"

The silvery ray answered, "I am rather special, or *critical* as you may say, but I do belong to one of the clubs. Guess, which?"

"Think Alice, think," prodded Dr. Whatif.

Alice saw the light. (There was quite a lot of light to be seen.) "If the critical ray was to meet the other side of the table in a point, I could find on the other edge a point *past* that meeting point, which could be joined to my *T*, to the point where the other red rays originate. So the critical ray would have red rays on its right *and* left, but then it could not be critical!"

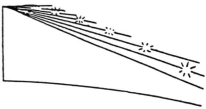

"You belong to the parallel club," stated Alice.

"I do, but I am a little different from them."

"In what way?"

Suddenly, on each of the blue rays a brilliant point lit up and then the lights went out just as suddenly as they appeared.

"What was that?" asked Alice. "What were those flashing lights on the parallel rays?"

"On each of the parallel rays the flashing light signified the point nearest the opposite edge," answered Dr. Whatif.

"Nearest? I would have thought that since those rays are parallel to the opposite edge, they keep at the same distance from it."

"Not in the hyperbolic plane."

"Oh?"

"You see, in a hyperbolic plane you cannot draw more than *one* common perpendicular to the two parallel lies."

Alice tried to make a sketch.

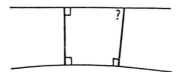

"I suppose," she said reluctantly, "in a place where rectangles are banned."

"You are getting there," said Whatif. "Now, suppose that you have one line perpendicular to each of the two parallel lines and you go along *one* of the lines, pick another point on it and you find that the distance from that point to the other line is ... ."

"The distance of a point from a line .... Can you drop a unique perpendicular from that point to the other line in this hyperbolic plane?"

"You are right to ask such questions. The answer to this one is yes.

The difference between your Euclidean plane and the hyperbolic plane is caused by a single axiom. In the Euclidean plane the parallel to a line through an external point is unique, in the hyperbolic plane it is not. All the theorems which you have proved at school and can be proved *without using this axiom* are still valid. So we leave your question at that. I must admit however that it was a good question."

Alice thought to herself, "At least he gives me credit for something."

"I will leave it to you to prove that the further you move from the common perpendicular, the greater the distance from the point to the other line. So the common perpendicular supplies the shortest distance from one line to the other. For the critical parallel, however, there is no shortest distance!"

"What is so special about it?"

Suddenly the mysterious lights again filled half the table. Alice heard a lilting voice coming from the silver line.

"I am very special indeed. I belong to the club of the blue lines, yet I have red lines as *near to me as I wish*. You know that all of them actually hit the other edge, so I get *as near to that edge as I wish*."

"But you do not hit it."

Before Alice finished her sentence, the lights faded again. All that she could see now was Dr. Whatif's grin.

"Is he the Cheshire cat?" she thought.

"The critical line is a *converging parallel*," declared the voice belonging to the grin, "while the other parallels ... ."

"Keep their distance."

"Not exactly! As you have seen, they do *not* keep the same distance, they come to a point nearest to the other line, then they increase their distance. To distinguish them from the critical line, they are called *ultraparallels*; and to make things simple, only the *two critical (or converging) rays* to the right and left from a point P are called *parallels*."

"I just can't see, *straight lines* doing such things."

(an ultraparallel)   (a parallel)

"Her thinking is in a straitjacket," sneered the Hare.

"I have been taught straight thinking," retorted Alice.

"You are just straitlaced."

Dr. Whatif cut the argument short. "You cannot define what a straight line is. These lines play the rules by the hyperbolic axioms, which are just the same, but for one, as the Euclidean ones. They may not gratify your preconceived ideas, but should appeal to straight thinking."

"I have not seen anything *similar* before," protested Alice meekly.

"Hah, similar," laughed the Hare. "We do not know such a thing here!"

"He is not only rude, but ignorant too!" thought Alice. "You have never seen similar triangles?" she asked aloud.

Dr. Whatif took over again, "In this plane, we have *no* similar triangles."

"But surely, you can *draw* two triangles with equal angles."

"In that case, they are congruent. If they were not, and you will prove this in detail, you would end up with a quadrilateral with an angle sum of 360° when you try to fit the smaller triangle in the larger one. You know that you can't have *that* here!"

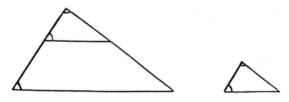

Words of dejection came to Alice's lips

I have been rather unprepared
For what I now detect,
Triangles and polygons
With angular defect,
And those converging parallels,
What an odd effect!

Then those ultraparallels
Spurting from point *P*!
Rays with ways so wild and weird,
Bewildering to me:
How can rays have turning points,
I find it hard to see.

Where are all the rectangles
Where are all the squares?
No more can the triangles
Form similar pairs.
Oh, it's a strange wonderland
Full of pits and snares.

"Now, now," said Dr. Whatif soothingly, "have a nice cup of tea!"

"At this table?"

However, to her delight, real cups of tea with cakes appeared in place of the T's. She followed the example of Dr. Whatif and gulped down her tea. It tasted good.

She looked around and noted that the Hare who did not touch his tea seemed to grow very big until he grew out of sight. Things became much more pleasant. She was sitting now at the side of a beautifully set table. Dr. Whatif was still there, but she was delighted to see Lewis Carroll's familiar face.

"What, what happened?"

"I saw that you needed a *shrink*," laughed Dr. Whatif. "The tea came from the shrinking bottle."

Alice played with the little *square* shaped mat at her setting. "You would not have a protractor by any chance?" she asked.

L.C. pulled one out of his pocket. Alice quickly measured the angles. They were all exactly 90°. She cried out triumphantly, "We are back at last in the Euclidean plane. Everything is as in the good old days."

"I do not know," cautioned Whatif. "Maybe we are, maybe we are not!"

"What do you mean?"

"It is possible that we are still in the hyperbolic plane."

"The square has no angular defect!" (Alice was proud of her new knowledge.)

"Perhaps it has, only it is so small that even your extraordinary eyes cannot measure it. It is possible that the *angle of parallelism* is very nearly 90°, whereas . . . ."

"Angle of parallelism?" interjected Lewis Carroll.

"Has it got anything to do with the silver ray?" asked Alice.

"Perhaps we'd better tell L.C. about the silver ray, he was not there, you know."

"We looked at a line ℓ, a point P outside it and looked at the rays coming out of P and not intersecting line ℓ. The least value of the angle that a parallel ray can make with PQ, the perpendicular to ℓ, is called the angle of parallelism."

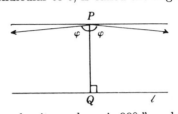

"Least value? The *only* value it can have is 90°," exclaimed Lewis Carroll.

"Yes, if you stick to your Euclidean parallel postulate. But taking the hyperbolic axiom instead, and allowing more than one line parallel to ℓ at P, you will get on each side of PQ a critical, or limiting parallel line, enclosing an angle ϕ with PQ. This is the angle of parallelism that we are speaking about," countered Whatif.

"But," started Lewis Carroll.

"I admit that it is very close to 90° here."

"Why not *exactly* 90°?" insisted L.C.

"The value depends on p, the length of the segment PQ. It is exactly 90°, only if p is exactly 0."

"If we have a hyperbolic plane," added Alice knowledgeably. (She enjoyed her advantage over L.C.)

Whatif continued, "The situation is a little similar to what you find if you look at your lawn. You treat it as a horizontal plane, part of your infinite Euclidean plane, whereas in reality it is a small spherical segment of our globe. If you tried to draw lines on a tennis ball, you would discard plane geometry pretty quickly!"

"So what about that 'angle of parallelism'?" queried L.C.

"It is easy to show that it cannot increase as p gets larger."

"I believe that! It always remains 90°," said Lewis Carroll, still sceptical.

"But if the plane is hyperbolic, it can be shown that for a suitable value of p, the angle of parallelism can be made smaller than any prescribed acute angle! In fact, with a little help, Alice will prove it later."

"All this means," said Alice eagerly, "that the angle of parallelism decreases from 90° to 0 as p increases."

"Well, our arguments make it plausible. In fact, Lobachevski ... ."

"Who?" asked Lewis Carroll.

"The poor fellow died, before he could make you take him seriously. Lobachevski found the exact relationship between p and the angle of parallelism. The angle is 90°, only if $p = 0$ and decreases towards 0° as p increases."

"If it is the 'hyperbolic plane', " countered Lewis Carroll.

"If it is *a* hyperbolic plane."

"So now you have many sorts of this thing," laughed Lewis Carroll.

"You can have spheres with different diameters. You can have different hyperbolic planes. A hyperbolic plane is fully determined by the value of p, for which the angle of parallelism is 45°," further explained Whatif.

"I would like to know just how far you have to go to achieve 0° as your angle of parallelism? I cannot see that it *ever* becomes different from 90°."

"Because," said Whatif, "all the triangles that you can see here are of a very small *area*."

L.C. raised his eyebrows, but Whatif continued, "Of course, one must realise that the area of a triangle may become large, but never reaches a certain upper bound however large its sides become."

This was too much for both Lewis Carroll and Alice. After her experiences, Alice could boast some expertise.

"Speaking about areas! There are no rectangles in the hyperbolic plane, let alone squares, like the square inch or the square mile for measuring areas."

L.C. added, "upper limit for areas! There are limits to my . . . ."

"Imagination?" laughed Whatif. "There is another Jabberwock for you!"

With the suddenness to which Alice had become accustomed, he produced a globe. Only this time there were no mountains and rivers, or continents and countries on it. Only the North Pole and the Equator and a few North-South lines were marked. He turned to Alice.

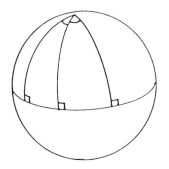

"We have no squares on the sphere either, have we? However, you can see things looking like triangles on the surface. If someone gave you the value of S, the surface area of the hemisphere, you would not find it too hard to find the area of those spherical triangles that you see, eh?"

Alice hesitated for a moment. "I think, ..., I think, one of them would have $\alpha/360$ times the area S as its area, the area of the other one is $(\beta/360)S$." She added more confidently, "The two of them add up to $((\alpha + \beta)/360)S$, the area of the large triangle."

"Well, you have that Jabberwock almost slain by now!" congratulated Whatif and he continued, "each of those triangles that you see has two right angles, so the angle sums are $180° + \alpha$, $180° + \beta$, $180° + \alpha + \beta$. In any spherical triangle the angle sum *exceeds* $180°$. Quite the opposite to the triangles in the hyperbolic plane. You have expressed the areas quite correctly, in terms of the spherical excesses, α, β, $\alpha + \beta$. You could choose $S/360$ as your area unit and could say that α, β, $\alpha + \beta$ are the measures of those areas. In fact, even if the triangles are more general, the spherical excess still gives the measure of the area."

"What has this got to do with the hyperbolic plane?"

"It is just the other side of the coin. In the hyperbolic plane, it is the angular *defect* that measures the area of a triangle."

"The sphere is all right, but why should this be so for the hyperbolic plane?" asked Lewis Carroll.

"The angular excess on the sphere and the angular defect in the hyperbolic plane both have the properties desired for measuring areas."

"What properties?" wondered Alice.

"First, the areas of two congruent triangles must be equal. Second, if a triangle is partitioned into two other triangles, then the area of the whole should be the sum of the areas of the two parts. It is easy to see that the excess or the defect satisfies these conditions."

"But is this sufficient?" asked Alice.

Lewis Carroll added, "You should also show that if the triangles have the same angular defect, then they can be partitioned into congruent parts."

"That can also be shown. It takes a little time. You should have read the proof that Bolyai has for it."

"If you *believe* in triangles having an angular defect!" L.C. could not be convinced.

"That is up to you. However, in that case, you must see that the area cannot exceed a certain value."

"Can the defects add up to $180°$?" asked Alice.

"If they do, there is no triangle!" answered Lewis Carroll.

"So here you have an upper *limit* for your area," added Whatif with a note of finality.

"And the upper bound of our credulity," added L.C.

Dr. Whatif burst into rhyme.

> Father Christmas brought to Jimmy
> Two dolls in a pack.
> Both of them had grinning faces,
> Both of them were black,
> One of them was Joey doll,
> The other was named Jack.

Both of them wore shiny garments,
Neither had a crack.
Yet Jimmy cuddled Joey doll
And angrily punched Jack,
And flung him nearly down the river
With a mighty smack.

Mother gathered up the poor doll,
And gave Jimmy a whack.
She put the two dolls tight together,
And tied them in a sack.
—Now you will lose Joey doll too,
If you toss out Jack!

"And what has Black Jack got to do with the angle of parallelism?" asked L.C.

"That is easier to answer than your puzzles. The Euclidean plane and the hyperbolic plane are both *extensions* of physical reality. Both are built on systems of axioms, in fact they differ in one axiom only, and neither structure shows any cracks."

"Euclidean geometry has had a much longer testing time."

"Admitted. I shall show next that the two alternative geometries can be tied together in such a way that if one of them is consistent, so is the other one. But this will be another story."

𝔓𝔯𝔬𝔟𝔩𝔢𝔪𝔰 𝔞𝔫𝔡 𝔈𝔵𝔢𝔯𝔠𝔦𝔰𝔢𝔰 4

Note. Those results of Euclidean geometry which can be established independently of the Euclidean parallel postulate, i.e., depending only on concepts of incidence, betweenness, congruence, and continuity, are valid also in hyperbolic geometry. A systematic, complete treatment of these can be found in a number of books. (See the bibliography and the list of axioms at the end of the book.)   Here is a short list of results (independent of the parallel postulate) that may be of use in the exercises to follow.

a. *Congruence* of line segments, angles (right angles in particular), triangles (SAS, SAA, ASA, SSS).

b. *Half planes*: A line ℓ divides the points of a plane into two disjoint sets, called half planes. A line joining two points belonging to different half planes intersects ℓ.

c. *Inequalities*

  i. The sum of the lengths of any two sides of a triangle is greater than the length of the third side.

  ii. An exterior angle of a triangle is greater than any of the two nonadjacent interior angles.

  iii. If two angles of a triangle are unequal then the side opposite the greater angle is longer than that opposite the smaller one.

  iv. If two sides of one triangle are equal in length, respectively, to two sides of another triangle, and the angle included by the two sides of the first triangle is larger than the corresponding angle in the second triangle, then the third side of the first triangle is longer than the third side of the second one.

1. Prove the *converse* of the Euclidean parallel postulate. It can be stated as follows. Let ℓ and m be two lines and let a transversal line intersect them at the points P and Q respectively. Let L, M be two points on ℓ and m respectively, on opposite sides of the transversal PQ. Let the alternate angles LPQ and MQP be equal. Show that the lines ℓ and m do not intersect. Hence show that if the *corresponding* angles RPL and PQM' are equal (where R is a point on QP and M' a point on MQ), then the lines ℓ and m are parallel.

Hint. Suppose that X is a point on PL such that X is also on line m. Find a point Y on QM such that the segments PX and QY are equal.

2. Prove (without referring to the parallel postulate) that the sum of the angles of a triangle cannot *exceed* 180°.

Hint. Assume that the sum exceeds 180°. Referring to the diagram in the text, denote the lengths of the segments $A_1 A_2$, $B_1 B_2$, $A_1 B_1$, $A_1 B_2$ by a, b, c, d, respectively. Show

    i. that $a < b$;

    ii. that for any positive integer value of n, $c + d + na > (n+1)b$.

Hence find a contradiction.

3. Let ℓ be a line, P a point not on ℓ, and Q the foot of the perpendicular from P to ℓ.

    i. Let A and B be two points on ℓ such that A is between Q and B and the segments PA and AB are equal in length. Show that the angle PBQ does not exceed half the magnitude of the angle PAQ.

    ii. Show that for a suitable choice of R on the line ℓ the angle PRQ can be made as small as we wish.

    iii. Suppose that the parallel postulate of Euclid does not hold for the line ℓ and the point P, i.e., there are at least two lines through P parallel to ℓ. Show that then there exists a triangle with angle sum less than 180°. (Thus, by the arguments in the text, every triangle in the plane has an angular defect different from zero.)

4. Let ℓ, P, and Q be as in 3. Denote the length of the line segment PQ by p. Show that in a given hyperbolic plane

    i. the angle of parallelism depends only on the value p, i.e., it is independent of the positions of ℓ and P;

    ii. for all values of p different from 0 the angle of parallelism is less than 90°;

    iii. as p increases, the angle of parallelism cannot increase.

5.

    i. Show that the angular defect is additive. This means that if a triangle is partitioned into two triangles, then its angular defect is the sum of the angular defects of the part triangles. (This can be generalised to any number of parts.)

    ii. Show that in a given hyperbolic plane the angle of parallelism can be made as small as desired by choosing p, as defined in 4, suitably large.

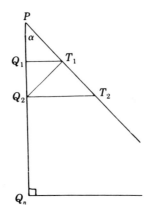

Hint. Starting with P and a given arbitrary acute angle α pick an arbitrary point T_1 on one arm of the angle and draw T_1Q_1 perpendicular to the other arm. Let Q_2 be a point on PQ_1 produced such that the segments PQ_1 and Q_1Q_2 are equal. Draw Q_2T_2 perpendicular to PQ_2 with T_2 on PT_1. (Is this construction always possible?) Continue by finding Q_3 on PQ_2 produced such that $PQ_2 = Q_2Q_3$ and draw Q_3T_3 perpendicular to PQ_3 with T_3 on PT_1. Repeat the procedure as long as possible. (Use the additivity of the defect.)

6. A quadrilateral $ABCD$ (named the Saccheri quadrilateral) has BC as a base, right angles at B and C, and the sides AB and CD are equal in length. The midpoints of BC and AD are E and F respectively. Show that in the *hyperbolic plane*

  i. the line EF is perpendicular to both BC and AD and hence AD is
     parallel to BC (This is also true in the Euclidean plane.) ;

  ii. the angles at A and D are both equal and acute;

  iii. the length of AD is greater than that of BC.

7. In the hyperbolic plane, ℓ and m are two parallel lines with a common perpendicular LM, where L is on ℓ, M is on m. A point Q moves along the line m. Show that the distance of Q from ℓ is minimal when Q coincides with M and that the further Q moves from M, the greater the distance of Q from ℓ becomes.

Hint. Let P be the foot of the perpendicular from Q to ℓ. Construct a Saccheri quadrilateral having vertices at L, P, and M.

8. Show that if two triangles in the hyperbolic plane are equiangular, then they are congruent.

9.

  i. N and S are two diametrically opposite points on the surface of a sphere.
     Two great circles (i.e., circles obtained on the surface by intersecting the

sphere by planes through its centre) intersect at angle α degrees at N and S. They divide the surface of the sphere into four regions, called *lunes*. Express the area of the shaded lune in terms of A, the surface area of the sphere and the angle α. (See the figure on the left below.)

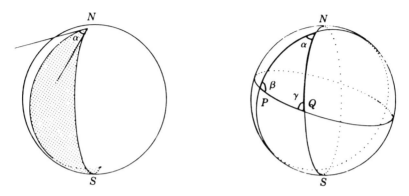

ii. A third great circle is drawn to intersect the bounding circles of the shaded lune at P and Q, at angles β and γ degrees, respectively. The surface of the sphere is now divided into eight regions called *spherical triangles*. Show that the area of the spherical triangle NPQ (as shown on the diagram) is $\dfrac{\varepsilon}{180} \cdot \dfrac{A}{4}$ where $\varepsilon = \alpha + \beta + \gamma - 180$. ε is called the *spherical excess*. (See the figure on the right above.)

Hint. Consider the areas of the *supplementary* spherical triangles, i.e., triangles which complete the triangle NPQ into lunes of α, β, γ degrees respectively.

Chapter 5

Circle—Land Revisited

$\mathfrak{T}$HE time has come," friend Whatif said,
"To talk of many things,
Of points and lines and congruence
Of groups and fields and rings."

"Oh no," said Lewis Carroll, "that is not what you said in the preface. After all, our young Alice is just out of finishing school! You cannot go too far with her with that twentieth century stuff."

The three of them were walking along a road, signposted 'To Geometry.' They had just arrived at a fork. Two arrows pointed to different routes: 'Algebraic' and 'Elementary.' "As you wish," answered Whatif to Lewis Carroll's protestation, "however, if we take the 'Elementary' route, we will not get quite so far."

Alice felt a little hurt. "I have been out of elementary school for a long time!"

"You should not be quite so disdainful towards that word: element. I have promised to tie the geometry you have newly glimpsed to the geometry of Euclid you both regard with great respect. The title of Euclid's work is *Elements*. You should really feel in your element."

With that remark Dr. Whatif turned to the right and motioned his party to go with him. The road they chose seemed to be pretty old, but it led through a landscape that was by no means dull. There was lush growth everywhere, with

83

exciting little roads and pretty little nooks. Figures well known and loved by Alice decorated the paths. Whatif chose a path that led towards a large circular structure, built in an impressive style. Catching a view of it, Alice became quite excited and began to run in its direction.

"Not so fast, young lady," said Dr. Whatif, pointing to a bench, "let us rest here for a while and prepare for our entry to the great circular hall."

"The circular hall? I think we have already done some preparation," said Alice.

"Sure. I hope that orthogonal circles and inversion are still fresh in your mind but

> The time has come, I say again
> To talk of many things,
> Of points and lines and congruence
> And groups ..."

"We draw the *line* here," interrupted Lewis Carroll, "we have agreed."

"All right, I get your *point*," said Whatif, "but we have also agreed not to *define* what *points* and *lines* are."

"What about congruence?"

"Will also remain undefined."

"Undefined?" asked Alice. "Can't we just say that the two figures are congruent if we can bring one to cover the other?"

"Oh, there we go! Alice, do you remember your ruler turning into a caterpillar when you moved it? How can you be sure you can move a line segment or an angle or a triangle from one place to another without changing it?"

"They do not seem to change very much!"

"Still, if we strive for precision, we cannot neglect the possibility."

"That makes it difficult," objected Alice.

"Not at all! We simply say: never mind what the words 'point,' 'line,' and 'congruence' mean. Let us *assume* that they play our game."

"What?"

"Fit our ideas about them. They are subjected to some basic rules we call axioms. We can go on from there."

He suddenly produced an odd looking package. On first glance Alice thought it was a pair of Siamese twin dolls (maybe Joe and Jack tied together), but as Whatif turned it, she saw that it was:

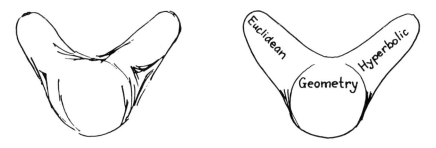

"Without *defining* the words," Whatif continued, "'point,' 'line,' 'on it,' 'between,' 'congruent,' and 'continuous,' and using for our base a few *axioms* involving these undefined concepts you can build up either the Euclidean or the hyperbolic geometry. They differ only in that *one* axiom."

"But the two geometries contradict each other," countered L.C., "so you must make your choice between the two!"

"I choose Euclid," said Alice.

"In that beautiful circular building you will come to see that the two alternatives, far from contradicting each other, are tied together," said Whatif.

"Just like Joe and Jack?" asked Alice.

"In a way, yes. You see, the geometry that you know 'by experience' does not tell you whether that disputed angle of parallelism is equal to 90°, or takes values very nearly equal to 90° at distances within the range of our measuring instruments. You could ask:

What if the angle of parallelism is 90°?
or
What if the angle of parallelism is different from 90°?

You get a perfectly good geometry in either case."

"I believe the Euclidean variety is 'perfectly good', but the other one has not been with us long enough to test all its conclusions," was Lewis Carroll's opinion.

"We will never be able to test *all* the conclusions of Euclidean geometry either. But, so far, so good. All we can say with *certainty* is that *if* Euclidean geometry is free of contradiction, then so is hyperbolic geometry. Jack and Joe thrive or wilt together."

"How can you be so certain, Dr. Whatif?" asked Alice.

"You will be just as certain after our visit to our 'model' in the circular hall. All you will find there is Euclidean geometry! For a Victorian young lady," he admitted reluctantly, "you seem to manage the Euclidean stuff reasonably well."

"But I have no grudge against Euclidean geometry."

"I know that. However, you will find that having freed ourselves from the slavery of definitions, you could also view the scenery as a hyperbolic geometry."

"I cannot say that you have made yourself clear to 'a Victorian young lady' or anybody else for that matter," quipped L.C.

"Concepts, such as 'point,' 'line,' and 'congruence' are left as blanks. As long as our axioms are satisfied, we can fill in the blanks any way we like. Why, we have already had an excursion to a place where lines were circles and congruence was induced by reflections on these lines."

"You mean inversions?" asked Alice.

"Yes."

"I could see that inversions behaved a *bit* like reflections, but ... ."

"It seems we must have some reflections on this matter.

> In geometry, you must all see
> We're striving for precision,
> Physical "realities" must
> Not obstruct our vision.
> Sweep aside all prejudice,
> Every fuzzy notion.
> Let us not talk any more
> Of things like 'rigid motion.'
> (Why, that thing is dead and gone
> With ancient generations!)
> Nowadays, we are dealing with
> *Groups of transformations.*"

"I wish he could curb his mania for rhymes," thought Alice, but did not say it aloud. Fortunately, he continued in prose.

"I have listed in your homework package a modern version of the axioms on congruence. From these, together with the other axioms, excluding the one on the parallels, all theorems common to Euclidean and hyperbolic geometry can be deduced. Now let us see the connection between reflection and congruence."

"What exactly do *you* call a reflection?" asked Alice.

"I have been waiting for that question. A reflection is a transformation."

"What is a transformation?"

"I should have started there. A transformation assigns to every point of the plane

an image point. The transformations we are interested in now are such that
(a) every point in the plane is the image of one and only one point, and
(b) the images of all the points on a
line lie again on a line."

It was still a little strange to Alice. She
drew a figure on the ground. "For me a
reflection in a line is just like looking in a
plane mirror. B is the image of A in line
ℓ if ℓ is the perpendicular bisector of the
segment AB."

"No objections. I want you to see that reflection is a very special kind of transformation, subject to rules (a) and (b) and furthermore subject to
(c) points on the *axis of reflection* ℓ, remain unchanged,
(d) the image of a point on a line m perpendicular to ℓ is also on line m and
(e) the angle between two lines remains unchanged on reflection, though the sense
of rotation becomes opposite."

"But the last one is the same as inversion in a circle" exclaimed Alice.

"That's it! Moreover, if you replace the words 'axis of reflection' by the words
'circle of inversion' and 'perpendicular line' by 'orthogonal circle', then you find
that all properties (a), (b), (c), (d), and (e) hold for inversion. Can you see it?"

"Yes," said Alice after a little thinking.

"This is very important," said Whatif standing up and resuming the walk towards
the circular building, "because the properties (a), (b), (c), (d), and (e) make sure
that the transformation is a reflection as you know it. You will not find it hard to
prove using your good old Euclidean congruence arguments."

Here Lewis Carroll interrupted. "What about the *identity* transformation? It
has properties (a), (b), (c), (d), and the angle remains unchanged, since everything
is unchanged."

"But the sense of rotation is reversed by reflection and remains the same under
identity. Anyway, I was thinking of *proper* transformations, excluding the identity."

They were approaching the building.

"Before we enter," said Whatif, "I want to say a word or two about congruence.
Earlier, Alice spoke about physically *moving* one figure into a congruent figure, and
we raised some objections. However, we can now replace that idea by something
more subtle. Under reflection in a line the image of a line segment is a congruent
line segment and the image of a triangle is a congruent triangle, though the sense
of rotation in which the corresponding vertices follow each other is reversed."

"Obvious," noted Alice.

"Well, you will have to prove it later, but it will not be hard. Moreover, two
consecutive reflections about two different axes will produce congruent figures with
the sense of rotation the same. So you could call such a pair of reflections a
movement of the plane."

"And if the axes of reflection are the same?" teased Lewis Carroll, turning to

Alice.

"We will be back where we started. That happens to me sometimes," she added, thinking of past Wonderland wanderings.

"We have then the identity transformation, the 'staying still' movement. Just a special case of a congruence transformation," said Whatif.

"I hope I am not expected to *prove* that a figure is congruent to itself!"

"Read your axioms. However, there will be a bit more for you to prove. First, you will be able to show that if two line *segments* are congruent, then you can get from one to the other in not more than *two* reflections and then you can proceed to prove that if two *triangles* are congruent, then you need at most *three* reflections to match them."

They arrived now at the building. Whatif announced with finality, "This is the way we look at congruence. *Two configurations are congruent if and only if we can find a finite number of successive reflections to transform one figure into the other.*"

They entered the building. They found themselves inside a spacious circular hall. Shining circular arcs lit up whenever Alice glanced at the floor. She suddenly had a feeling of recognition, a vague sort of familiarity. It was somewhat like that place they had already visited with Dr. Whatif. ("There are shining circles all over the floor, and yet .... .") Just then two striking looking hostesses approached them, one from each side.

"Welcome to Poincaré Hall," said the one, dressed all in white "I hope that you will feel comfortable here. It is all Euclidean, you know," she added, turning to Lewis Carroll.

"I must warn you, however," said the other hostess, wearing flowing red garments, "that our boundary, the great **P** circle, is strictly out of bounds." Her voice sounded

rather stern as she pointed at the silver circle around the internal walls of the building.

"Is she the Red Queen?" thought Alice somewhat uneasily.

The White Hostess laughed, "There is not much danger that anyone within this hall can reach Infinity."

This was a little puzzling to Alice. Her bewildered expression invited a severe glance from the Red Hostess.

"We do not ask for passports here and we do not charge admission fees, but we must make sure that our visitors are well versed in Inversion!"

"Surely, you will accept Dr. Whatif's and Lewis Carroll's credentials," said the White Hostess.

The Red Hostess turned to Alice. "Do *you* know what inversion is all about?"

"Yes," said Alice, eager to show her competence. "I can construct the inverse of a point in a circle." She added, "it is just like reflection."

"And what *right* have you got to state that it is just like reflection?"

Alice was frightened by the sharp voice of the Red Hostess. She looked at Lewis Carroll but, before he could help, Dr. Whatif interrupted impatiently, "we discussed it before entering."

Alice gathered her courage. "It is a transformation, having all the properties of reflection, if we regard the circle of inversion as the axis of transformation."

The White Hostess looked more friendly. She motioned Alice to go with her and look at the shiny circles on the floor. "Of course, you cannot see *all* of them at once. They cover the floor densely everywhere, in fact, an infinite number of them go through every point."

"Points! You'd better point out that what *we* call points are strictly *inside* the **P** circle." The Red Hostess sounded rather *pointed*.

"Yes, in our geometry we have no points outside. Even the points on the **P** circle are out of bounds. We call them *ideal* points."

"This is a little worse than what we saw in Dr. Whatif's circleland. He wiped out only one point," thought Alice.

Aloud she asked, "And the circles? Does every circle through an inside point belong to this ... establishment?"

"Geometry!" corrected Dr. Whatif.

"Oh, no," answered the White Hostess. "You must notice that all *our* circles are *orthogonal* to the **P** circle."

Indeed, as Alice focused her eyes, she noted that each of the shining circular arcs hit the **P** line at right angles. She noticed something else. "Some of these circles are really straight lines!"

The Red Hostess summoned Alice to the centre of the hall.

"Have you ever tried to find a circle orthogonal to a given circle and going through the centre of the circle?"

Her sarcastic tone put Alice off. She remembered, "I could construct through two given points a unique circle orthogonal to a given circle, unless ... ."

Dr. Whatif came to the rescue. "Circle or straight line, it really doesn't make much difference to us since we invited infinity to our inversive plane, however ... ."

"Perhaps, she could have something about this in her problem package," said Lewis Carroll.

"Yes. She will be able to prove that orthogonal circles become diameters if they pass through the centre of the given circle."

The White Hostess produced three pairs of delicate opera glasses and handed them to the visitors.

"These are our 'logical glasses.' If you look through them you will be able to see Poincaré Hall as an infinite plane but make sure that you disregard all points not inside the **P** line. The circles orthogonal to **P** will appear to you as *lines*."

They put the glasses to their eyes. Lewis Carroll said, "There is nothing surprising about all this. The glasses make you concentrate on a small region, which is, of course, a good approximation."

"But when you remove your glasses, you can still use Euclidean geometry to *prove* your observations," said Whatif.

"Which observations?" asked Alice. She removed her glasses and added, "I observed that *two lines meet in at most one point*, but now I see them for the circles they are, and I know that two circles meet in two points."

"Oh," laughed the White Hostess.

"Oh," snarled the Red Hostess.

"You have forgotten the rules, Alice," reminded Dr. Whatif. "Points outside the

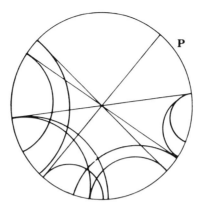

Hall do not exist. The other intersection of two orthogonal circles through some point P is the *inverse* of P about the **P** line, so it must be *outside* and so it *does not exist* in our geometry. So the situation is the same as in the geometry you learned earlier: at most one intersection point. You can discover other similar features. Each line ℓ ....”

“Circular arc?”

“Yes, circular arc, generally, but it could also be a diameter, since it divides this circular hall, our plane, into two half planes. If a point A belongs to one, and a point B to the other half plane, then the line joining A and B must intersect ℓ. Moreover, in the Euclidean plane, as you know, a reflection about a line takes all the points of one half plane to the other half plane.”

“Can we be sure that inversion about ℓ takes *all* the points of one half (it does not always *look* like one *half* to me) into the other half? Will all the images stay *inside*?” asked Alice.

“You should remember what you already know about inversion.” Dr. Whatif made a sketch of the situation, with a circle **C** playing the part of **P** but, of course, in that sketch points external to **C** did exist.

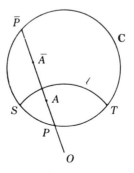

“If you know that the circle **C** is orthogonal to circle ℓ, what happens to its inverse in ℓ?” It was Whatif’s turn to ask a question.

"**C** goes into *itself*. The intersection of ℓ and **C**, the points S and T stay still, and the arcs between them are *exchanged*," Alice said promptly.

"Top mark! So a point P on **C** goes to a point $\overline{P}$ on the opposite arc and any point A *between* P and ℓ on the segment joining O (the centre of the inverting circle) and A goes into a point $\overline{A}$ *between* ℓ and $\overline{P}$. No danger of escaping outside!"

"Earlier you spoke of *the* line joining A and B. Is there a unique line?"

"You should remember that two lines meet in one point only, so you cannot have *two* lines through two distinct points A and B." This was L.C.'s remark.

"It does not follow that we *can* have *one* line!" answered Alice.

"Not bad thinking," said Dr. Whatif, showing his approval "but you should remember that if you count diameter lines as special kinds of circles then you can always find through two given points A and B a circle orthogonal to a given circle."

"I know what is coming next" said Alice. "You have talked about incidence and intersections and now we are coming to congruence."

The White Hostess handed the 'logical glasses' back to Alice with an approving smile. She stretched herself out on the floor and pointed to one of the shiny circular arcs.

"Can you invert me through this?"

Alice performed the construction just as she learned it and then looked through the glasses. The inverted image was just as tall and beautiful as the White Hostess herself, distinguishable from the original only by the exchange of right and left, exactly like a mirror image.

"But through these logical glasses I do not see things as they *really* are!" exclaimed Alice and removed the glasses. The lines became circles again and the image of the beautiful Hostess became a small crumpled heap.

"If I could only reduce the Red Hostess to size," thought Alice wistfully.

Whatif was quite ready to deal with Alice's objections. "In the geometry of this Hall the things you call points and lines satisfy the same theorems of congruence as you are used to in your old geometry."

"Only if you define congruence the new way," objected Lewis Carroll.

"We have agreed on this. We call two configurations congruent, ..."

"if one is the image of the other after some successive inversions," finished Alice, because she did not want to hear another lecture. She was not yet happy. "I can see that you can always make one point the inverted image of another point ...," pondered Alice.

"Well, this is true, but you will have to prove it by finding the axis of inversion," warned Whatif.

"Then," continued Alice, "in the Euclidean plane I can always copy a given line segment to any given line using any of its points as one extremity of the segment."

"You can do it here. Given two points A and B on ℓ, one of those arcs orthogonal to **P**, and a point P on another such arc m, you can find two points Q and R, to the right and left of P on m, such that the segments PQ and PR are congruent, in our sense, to the segment AB. It will not be hard to prove."

Alice thought for a while, then continued. "In geometry, we always *measure* things. You could put a ruler on a line and then measure out on the line as many inches, or other units, as you wish. You could go on and on. Moreover, if you went on long enough, inch by inch, then you could pass any point on the line." She bent down to one of the arches and began putting down little shiny markers, handed to her by the friendly White Hostess.

"I take A_1 and A_2 on ℓ and try to get points A_3, A_4, and so on, so that the arc segments A_1A_2, A_2A_3, A_3A_4 are congruent."

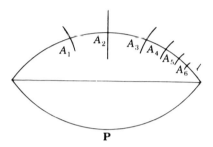

"Can she *do* this? asked the Red Hostess in a doubtful tone.

"Yes, I can," said Alice defiantly. She had regained her confidence by then.

"How?"

"Through A_2 I find the axis of inversion, orthogonal to ℓ,"

"And to $\mathbf{P}$," added Dr. Whatif. "You did prove a while ago that through each point of the plane, on ℓ or elsewhere, there is a unique circle (or line) orthogonal to both $\mathbf{P}$ and ℓ. Dropping perpendiculars to a line in this geometry works the same way as it does in your old geometry."

"I then invert A_1 to get A_3," continued Alice "and then invert A_2 about the axis through A_3, and so on, but .... ."

"But?"

"Aren't we bending the rule?"

"Bending the ruler, yes. Bending the rule, no. Everything will be just like in the Euclidean plane, up to the parallel postulate," assured Whatif.

"We have *our own* parallel postulate," snapped the Red Hostess.

"We'll come to that later," said Whatif. "First let us settle all your objections, Alice. What are they?"

"Well, this is a nice big circular hall, but after a while our construction will lead us to the *end* of arc ℓ or maybe past it."

"No danger of that, Alice," laughed the Red Hostess, "continue your measurements!"

Alice kept putting little markers on the lines. There were by now many markers, getting closer and closer. She looked at that arc—that Poincaré line—with all those little markers on it. She was not wearing her logical glasses. Her eyes (though extraordinary) were hurting, and her back was aching. She stood up, stretching herself. "Ugh, I have been working so long and I have covered such a small *distance!*"

"Hah! *Distance!*," guffawed the Red Hostess. She added,

> "Lose your breath! That is your lot,
> to stay stuck to the same old spot."

"Disaster," thought Alice, "she must be the Red Queen!" Now the painful experience came back to her, when the Red Queen had made her run faster and faster just to keep in the same place.

"But this is different," she reassured herself, as the White Hostess handed the logical glasses back to her. Looking through them, it seemed to her that her two companions were by now very far from her, somewhere near the centre of the hall. Yet, when she removed the glasses, they were near again. In fact, Whatif approached her, touching her shoulder.

"Here comes the lecture," thought Alice.

Indeed, Whatif produced a large board with a figure on it. As he was holding up the board, he announced, "You should be ready by now to learn about Poincaré distance."

"The circle **P** represents the boundary of our hall."

"All those *ideal* points?"

"Yes, but the *picture on this board is just an ordinary circle.* Just use your old Euclidean geometry, the way you learned it."

"With the new bits you have learned about inversion," added Lewis Carroll.

"I have made it simple for you," continued Whatif. "Since inversions preserve

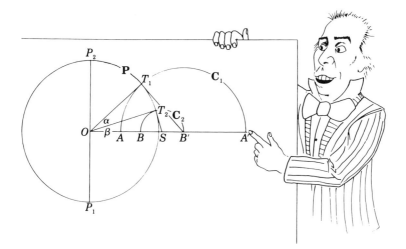

congruence in this Poincaré geometry, you may as well consider a walk along a *line* through O."

"A diameter?"

"Yes, making A your first stopping place."

"There are two more circles on the diagram, through A and B."

"Is there anything you notice about them?"

"They are both *orthogonal* to $\mathbf{P}$ and also to that diameter at A and B."

"Your eyesight still works very well."

"They both represent Poincaré lines."

"Your brain also works well," encouraged the White Hostess.

"Does it?" asked the Red Hostess. "How about the points A' and B'? Your Poincaré lines intersect the line OAB again."

But Alice saw the trap. "A' and B' are not Poincaré points. They are external to circle $\mathbf{P}$."

She continued hastily, before the Red Hostess could interrupt, "they are just *ordinary* Euclidean points on Dr. Whatif's board and they are inverses in $\mathbf{P}$ of A and B."

"With your *extraordinary eyesight*," teased Dr. Whatif, "you should have also discovered something else."

"I can see that B' is at the *centre* of the larger circle $\mathbf{C}_1$ ...," Alice stopped, thinking.

"Perhaps I was not fair," said Whatif, "I should have told you that B *is the inverse of O in the circle $\mathbf{C}_1$.*"

"She could have guessed it," countered the Red Hostess, "after all of those inversions she has been doing along the line ℓ."

Alice was not to be put down. She was thinking fast, to meet the challenge. "The smaller circle C_2 passes through the centre of C_1, hence its inverse in C_1 must be a line perpendicular to its diameter BB'. Now B and O are inverse in C_1, hence the *inverse of* C_2 *in* C_1 must be *the line* P_1OP_2 which is a diameter of P, perpendicular to OAB."

"Very good," said Whatif, "I hope you are just as good at *trigonometry* as at your Euclidean geometry. We will now need a little trigonometry."

Alice looked somewhat troubled.

"It will not be too bad. Let T_1 and T_2 be intersections of P with C_1 and C_2, respectively. Denote the angles AOT_1 and BOT_2 by α and β. Let the radii of the circles P, C_1, and C_2 be r, a, b, respectively. Now, where can you spot your right angled triangles in this picture?" asked Whatif.

"The centres of C_1 and C_2 are B' and S, respectively. Since the circles C_1 and C_2 intersect P orthogonally, the triangles $B'T_1O$ and ST_2O are right angled," Alice was quick to answer Whatif's question.

"So you can express a and b in terms of r, α, β:

$$a = r \tan \alpha \qquad \text{and} \qquad b = r \tan \beta.$$

What can you say about the (Euclidean) lengths of $B'B$ and $B'O$?" questioned Whatif again.

"Since B is the inverse of O in C_1 we have

$$B'B \cdot B'O = a^2.\text{"}$$

"Good," this time it was Lewis Carroll expressing approval. He helped Alice further. "Now express all those angles in terms of r and angles α and β."

"Since from right angled triangle $B'T_1O$ the length of $B'O$ is $\dfrac{r}{\cos \alpha}$ and $B'B = 2b = 2r \tan \beta$, we obtain

$$2r \tan \beta \, \frac{r}{\cos \alpha} = (r \tan \alpha)^2, \text{"}$$

answered Alice, after some thinking.

Whatif finished this calculation, "You can now work out β in terms of α and obtain

$$2 \tan \beta = \tan^2 \alpha \cos \alpha = \tan \alpha \sin \alpha.\text{"}$$

"This is not difficult," said Alice, "but you promised to tell me something about *Poincaré distance*. The way I look at it, especially through the logical glasses, B *should be twice as far from* O *in the Poincaré sense as* A *is, since it is the inverse of* O *in* C_1."

"That is just what we are going to discuss now. At this stage we use a trigonometric formula you may have learned and can remember, or you could look it up in a book. The three ratios $\sin\theta$, $\cos\theta$, $\tan\theta$, can all be expressed in terms of $t = \tan\frac{\theta}{2}$:

$$\sin\theta = \frac{2t}{1+t^2}, \qquad \cos\theta = \frac{1-t^2}{1+t^2}, \qquad \tan\theta = \frac{2t}{1-t^2}.$$

So if you put $t = \tan\frac{\alpha}{2}$ and $x = \tan\frac{\beta}{2}$, then after some simplification, the equation

$$2\tan\beta = \tan\alpha\sin\alpha$$

becomes

$$\frac{x}{1-x^2} = \frac{t^2}{1-t^4},"$$

"This means," cried out Alice, "that

$$x = t^2."$$

"The equation also has another solution: $x = \frac{-1}{t^2}$," cautioned Whatif, "but if you consider our diagram, then it is your solution that meets the situation."

Alice was not satisfied. "Dr. Whatif, you promised to show that the Poincaré distance of B from O is twice that of A."

"We are nearly there, Alice. I hope you remember what a *logarithm* is."

"Well, an exponent. The logarithm of 10 is 1, of 100 is 2, of 1000 is 3, and so on, meaning that $10 = 10^1$, $100 = 10^2$, $1000 = 10^3$."

"You should also consider

$$10^{-1} = 0.1, \qquad 10^{-2} = 0.01 \qquad \text{and} \qquad 10^{\frac{1}{2}} = \sqrt{10},$$

hence

$$\log 0.1 = -1, \qquad \log 0.01 = -2 \qquad \text{and} \qquad \log\sqrt{10} = \frac{1}{2},"$$

helped Lewis Carroll.

"Most importantly," said Whatif,

$$\log a^2 = 2\log a."$$

Alice now saw the light. "So, since we have

$$\log x = 2\log t,$$

could we call $\log t = \log\tan\frac{\alpha}{2}$ the Poincaré distance of A from O?"

"True, it satisfies the requirement that

$$\log\tan\frac{\beta}{2} = 2\log\tan\frac{\alpha}{2},$$

but it would not be quite *convenient*."

"Convenient?"

"What sort of angles are $\frac{\alpha}{2}$ and $\frac{\beta}{2}$?"

"Acute?"

"And not very large at that, less than ... ?"

"45 degrees," Alice was prompt.

"So $\tan \frac{\alpha}{2}$ is less than 1, for any choice of α in our diagram. That makes $\log \tan \frac{\alpha}{2}$ negative. You would not choose negative numbers for distances!"

"Not even in Poincaré country," laughed the Red Hostess.

Alice looked a little disappointed but help came quickly from her old friend, L.C., "Do you remember that $\cot \frac{\alpha}{2}$ and $\cot \frac{\beta}{2}$ are the reciprocals of $\tan \frac{\alpha}{2}$ and $\tan \frac{\beta}{2}$? Therefore, you can also write your equation as

$$\cot \frac{\beta}{2} = \left(\cot \frac{\alpha}{2} \right)^2 .$$

Then you deal with numbers greater than 1."

"So I could get the equation

$$\log \cot \frac{\beta}{2} = 2 \log \cot \frac{\alpha}{2} \ ?"$$

queried Alice.

"Or even more generally

$$c \log \cot \frac{\beta}{2} = 2c \log \cot \frac{\alpha}{2} ."$$

answered Whatif.

"What is c?" asked Alice.

"It is some conveniently chosen constant. You could then have the Poincaré distance p, corresponding to the angle α, be

$$p = c \log_{10} \cot \frac{\alpha}{2} ."$$

"Why could we not have $c = 1$?" asked Alice who was still a little puzzled.

"Writing c gives you more freedom in choosing your *unit lengths*. After all, once you are out of this hall, you can measure distances in feet or inches or miles, or even in centimetres or kilometres if you travel to Paris."

"So why not choose the unit length to make $c = 1$?"

"Your question is justified. The reason is that we want a different base for the logarithms."

"Base?" Alice was bewildered.

Lewis Carroll again came to her aid. "You have learned to take 10 as a base to make $\log 10 = 1$. But there is nothing special about the number 10.

For example, you could take 2 as the base, making $\log_2 2 = 1$, $\log_2 4 = 2$, $\log_2(\frac{1}{2}) = -1$, $\log_2 \sqrt{2} = \frac{1}{2}$."

"In fact," said Whatif, "we choose the number e as the base, to fix the unit for Poincaré-lengths and make

$$p = \log_e \cot \frac{\alpha}{2}\text{ ."}$$

Lewis Carroll intervened. "My dear friend, Whatif, our Alice is just out of finishing school. You cannot expect her to know about e."

"I apologise," said Whatif. "I trust that one day she will learn about it. It is a very important number, although not an integer like 2 or 10. It is called the *base of the natural logarithms* and is approximately equal to 2.718. Just as with base 10, the statement $\log_e A = \ell$ means that

$$A = e^\ell\text{ ."}$$

Alice looked a little happier. "I see now why you use log-ical glasses to measure distances in Poincaré land." Suddenly Alice turned to Whatif. She discarded the logical glasses and looked a little concerned. "I can see that the Poincaré distance OB is twice that of distance OA, but can we be sure that whenever *any two* segments are Poincaré congruent, their lengths measure the same? Can we be sure when we add *any two* segments, that the Poincaré length of the resulting segment is the sum of the lengths of the two?"

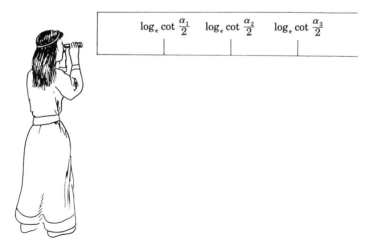

"She asks searching questions," noted the White Hostess with a smile.

"Indeed," agreed Whatif, "and I can give the answer *yes* to both questions. She will find in her problem package a little exercise which should make things clearer."

"It seems," said Alice, "that the Poincaré distances of A and B on the diagram depend only on those angles α and β. Have the angles any *special* name?"

"We will come to that," said Whatif, "and you should find the name quite familiar."

"I think, I am becoming quite familiar with a lot of things. All those *families* of circles, the circle **P** and the diameter holding A and B, they belong to an interesting elliptic family. The circles C_1 and C_2 and all of those others that I used for inversion, when I was measuring out congruent segments along the Poincaré line ℓ, are all orthogonal to **P** and line AB so they form a hyperbolic family."

Whatif nodded.

"And I now feel quite familiar with *congruence* in this hall, but .... ."

She looked down again at the shiny markers she had placed along line ℓ during her measuring efforts. "How can I be sure that the sequence of points $A_1, A_2, \ldots, A_n$ will get past any *internal* point on ℓ?"

"Now that is a good question. You will prove it later, with a little guidance. Leave it at that for now."

"How about SAS, SSS, ASA—all those congruent triangle theorems?"

"They are all valid in this geometry."

Dr. Whatif turned now to Lewis Carroll, who had been unusually quiet throughout this visit, feeling somewhat lost. Nobody could blame him for not having sense for nonsense, but his dividing line was strict. He could not tolerate nonsense where he believed the rule of sense was absolute. He treated Euclidean geometry with such reverence. Now he was confronted by something he could not classify as nonsense.

"We are now coming to the clinch. This beautiful model ("Is he talking about the White Hostess or the geometry?" wondered Alice) gives us a geometrical structure identical to the Euclidean with one important difference."

It was not hard to see that through any point P, not on the *line* ℓ, there were indeed many (infinitely many) circles orthogonal to the bounding circle **P**, not intersecting the arc ℓ.

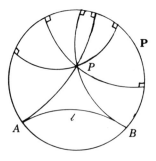

Alice remembered the *critical* parallel (the silver ray at the T-party). "I expect then that a critical parallel to ℓ through P is the arc through P that *touches* ℓ at A, the 'out of bound' point."

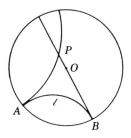

"Quite correct. In your earlier work you found the construction for it. It could be, of course, a diameter if P was a point on the radius OA."

"There is also a critical 'parallel' hitting $\mathbf{P}$ at B," said Alice.

"Sure."

"But they do not look 'symmetrical' to me," worried Alice.

"Symmetrical about what? Get your 'logical glasses' and think."

Alice looked again. "Yes, they do seem to make equal angles with the perpendicular through P."

"Be more precise," said the stern Red Hostess, as she approached the group, "what perpendicular do you speak about?"

"Why, the arc n through P, orthogonal to both ℓ and $\mathbf{P}$. We have settled that already."

"Of course, Alice will have to prove later that the perpendicular bisects the angle between the two critical arcs," said Lewis Carroll.

"She can prove it right now. I only have to remind her that P can be inverted to O, the centre of circle $\mathbf{P}$," ventured Whatif.

Alice took the hint. "All the orthogonal arcs through P become radii through O. The images of A and B are some points A' and B' on $\mathbf{P}$. The image of ℓ is some arc ℓ' orthogonal to $\mathbf{P}$. The image of the arc perpendicular to ℓ will be the radius n', still orthogonal to ℓ' and $\mathbf{P}$ since angles do not change on inversion. Since n' is a line through the centre of the circle $\mathbf{P}$ and also perpendicular to ℓ', it is the line joining the centres of two circles. Hence, it perpendicularly bisects the common chord $A'B'$ and so bisects the angle between the two radii OA' and OB'.

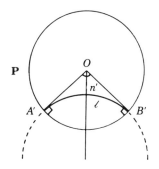

As all the angles remained unchanged through inversion, the arc n makes equal angles with the original arcs PA and PB."

Alice was pleased that the proof met general approval, even the Red Hostess seemed to be satisfied. So she carried on happily, "I think, we could use the same method to show that the triangles in this geometry have an *angular defect*. We invert again one vertex of any triangle to the centre O. Then two sides lie

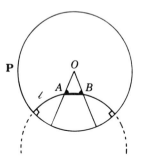

on radii of **P** and the third side is an arc ℓ *orthogonal* to **P**, so O is a point *external* to the circle to which arc ℓ belongs. If A and B are the remaining vertices of the inverted triangle, then it is clear that the angles are smaller than the angles between OA and the chord AB, and OB and the chord AB respectively. Hence the angle sum in the triangle formed by OA, OB and ℓ must be less than 180°."

"How about rectangles in this geometry?" Dr. Whatif was testing Alice.

"How could they exist if all the triangles are defective? The angles of two triangles could not add up to four right angles!"

"Do you remember that in hyperbolic geometry there was just one line perpendicular to each of two ultraparallel lines, ℓ and m?" asked Whatif.

Alice nodded.

"What is the situation here?" Whatif waited for Alice's answer.

"We could not have *two* common perpendiculars if there are *no rectangles*, but can we have *one*?"

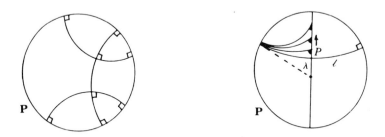

"Alice! You have not forgotten how to find a circle orthogonal to those given circles? Just look at that figure."

"How about that *angle of parallelism*?" It was Lewis Carroll who posed the question.

"No harm in choosing that arc perpendicular to ℓ along a diameter. Inversions can take you there," helped Dr. Whatif.

"It is easy to see that as soon as P leaves the arc ℓ, to move along the diameter, the angle of parallelism will become less than 90° and will get smaller and smaller," said Alice. "It is easy to see it in the figure."

"You will prove that later. You will also show that the angle of parallelism tends to zero as P approaches the end of the diameter."

As Alice marked the positions of successive 'critical parallels,' Whatif added, "it will also be interesting to construct the angle of parallelism at the centre O. Call it λ."

Alice thought for a moment before asking, "The critical parallel through O?"

"Well?"

"Why, it must be the radius OA! Of course, it must be orthogonal to **P**."

She completed the diagram, marking in the angle λ.

"Does this angle of parallelism look familiar to you?" asked Whatif.

"Oh," cried Alice, "it looks just like those angles α and β we had before on Dr. Whatif's drawing, when we talked about Poincaré distance."

"Does she still remember those formulae?" doubted the Red Hostess.

"I do! In this case I can say that the Poincaré distance ℓ from O ..."

"or from the diameter perpendicular to OP; think of that blackboard," helped Whatif,

$$\text{"is} \quad p = \log_e \cot \frac{\lambda}{2} \text{,"}$$

finished Alice.

"So now you can calculate exactly in this geometry the angle of parallelism, if you know the Poincaré distance p." Whatif turned to Lewis Carroll. "Once you create a geometry where you have more than one parallel through a point, you can get the *same* results you have found so unbelievable in the hyperbolic plane."

Lewis Carroll still kept quiet.

Whatif continued, addressing Lewis Carroll, "I can see that you accept what you see in this hall. After all, *this is Euclidean geometry*. But I have to point out again that by *our concept of congruence*, we created a *model* for *hyperbolic geometry*, the geometry of Bolyai and Lobachevsky, you have spoken so disparagingly about. In fact that formula for *Poincaré distance* or shall we call it simply 'inversive distance' is identical to the famous formula of Lobachevsky:

$$\cot \frac{\lambda}{2} = e^p$$

or

$$\tan \frac{\lambda}{2} = e^{-p}$$

the same as the Poincaré length we found before and whereby the angle of parallelism λ is calculated from the distance p."

Lewis Carroll looked pensive, and Alice found it a little hard to follow.

Whatif turned to Alice. "You will find everything more clear after you have made your way through your new problem package."

Alice was looking again at the shiny circles. "We regard all the circles that we can see here as lines. They all look *straight* through logical glasses. Are there any *circles* in this geometry?"

"That is a very interesting question. Your elliptic and hyperbolic families will give the answer. Take a family of circles, looking like lines through your logical glasses going through a point. They will be *one* family, your Montague's. Then the *corresponding* Capulets, the hyperbolic family orthogonal to it will give you as many circles as you wish. You will see it for yourself."

He turned to Lewis Carroll.

"I think we can stop here now. We could go on, having hours and hours of fun, checking hyperbolic theorems in this geometry, but I think that we have already reached the happy end."

"Who is to live happily ever after?" asked Alice.

"All of us in the knowledge that if we accept Euclidean geometry, we must accept hyperbolic geometry with it."

Lewis Carroll broke out in a broad smile.

> Euclid saw the light,
> Euclid was right!

"You are still not convinced, Uncle L.C.?" asked Alice.

"On the contrary, I am convinced now that Euclid was wiser than those generations of mathematicians who followed him and tried to *prove* the parallel axiom. He *knew* that it cannot be proved from the other postulates, because it is *independent* of them."

"They *should not call the hyperbolic geometry non-Euclidean!*" exclaimed Alice.

"Indeed, you are right," said Whatif.

Alice found herself holding the twins, Euclidean Joe and Euclidean Jack, in her arms. They were all *elated.* In a happy mood their feet left the ground and they were rising high above the Hall.

"It's just as well we can fly, I don't know how we could have gotten through that **P**-line which is Infinity in that Hall otherwise," thought Alice.

The two hostesses waved goodbye. The three of them flew up singing gaily.

𝕻roblems anð 𝕰xercises 5

1. Let ℓ be a line of the Euclidean plane. A reflection about ℓ assigns to each point P an image Q such that ℓ is the perpendicular bisector line of the segment PQ. Of the properties (a), (b), (c), (d), and (e) as listed on page 87, (b) and (e) require proofs. Prove—without applying the parallel axiom or its consequences (e.g., angle sum of triangles, parallel segment theorem) that

i. if A, B, C are points on a line, so are their images, hence the image of a line is a line,

ii. the angle between two lines remains unchanged on reflection, but the sense of rotation is reversed.

2. We are given that a certain transformation of the plane has properties (a), (b), (c), (d), and (e) as listed on page 87. Denote by P' the image of an arbitrary point P. Show that ℓ is the perpendicular bisector of PP', i.e., that the transformation is a reflection.

3.

i. Show that the image of a line segment on reflection in line ℓ is a congruent line segment, and the image of a triangle is a congruent triangle.

ii. Show that if two line segments are congruent, then one can be transformed into the other in at most two reflections and if two triangles are congruent, then one can be transformed into the other in not more than three consecutive reflections.

The following problems deal with the Poincaré geometry introduced in the preceding chapter. We use the following vocabulary:

Poincaré circle **P**: A circle of centre O. The Poincaré geometry is defined in a restricted region of the Euclidean plane, namely the interior of circle **P**.

Poincaré points : Points in the *interior* of **P**.

Poincaré lines: (a) Diameters of **P**, excluding the end points. (b) Arcs of circles orthogonal to **P**, consisting of the Poincaré points of those circles.

Ideal points: Points of **P**.

Poincaré reflection: (a) Reflection in a Poincaré line of type (a). (b) Inversion in a Poincaré line of type (b) (completed to a circle).

Poincaré congruence: Two configurations are said to be *Poincaré congruent,* if one can be made the image of the other in a finite number of successive Poincaré reflections.

4. Let A and B be two distinct Poincaré points. Construct the Poincaré line through A and B. (Recall Chapter 1, Problem 10.)

Discuss the questions of possibility and uniqueness of your construction.

5.

i. ST is the chord of intersection of two Euclidean circles $\mathbf{C}_1$ and $\mathbf{C}_2$ of centres O_1 and O_2, respectively. A and B are points on circle $\mathbf{C}_2$. Prove that the chords AB and ST of $\mathbf{C}_2$ are parallel, if and only if $O_1A = O_1B$.

ii. A and B are two Poincaré points. Find a Poincaré line such that B is the Poincaré reflection of A about that line. When is the axis of reflection of type (a)? Prove that a unique axis of Poincaré reflection exists in all cases.

6.

i. Show how to construct a circle orthogonal to a given (Poincaré) line and touching a given (Euclidean) line at a given point. Is the construction always possible?

ii. The Poincaré lines ℓ and m intersect at the Poincaré point A. Find two Poincaré lines such that each of them makes equal angles with ℓ and m.

iii. A and B are points on the Poincaré line ℓ. P is a point on the Poincaré line m (not necessarily different from ℓ). Find two points: Q and R on m such that the Poincaré segments PQ and PR are Poincaré congruent to the Poincaré segment AB.

7. Prove that Poincaré congruence satisfies the following axioms of congruence.

i. Addition of congruent segments: A, B, C are Poincaré points on the Poincaré line ℓ, B being *between* A and C. A', B', C' are on the Poincaré line ℓ', where B' is between A' and C'. Show that if the segments AB and $A'B'$ are Poincaré congruent, and the segments BC and $B'C'$ are Poincaré congruent, then so are AC and $A'C'$.

ii. SAS triangle congruence: Two Poincaré triangles ABC and $A'B'C'$ are such that the segments AB and $A'B'$ are Poincaré congruent, and so are the segments AC and $A'C'$. The angles between the arcs AB and AC, and the arcs $A'B'$ and $A'C'$ respectively are α and α' where $\alpha = \alpha'$. Show that the two triangles are Poincaré congruent.

8.

   i. D is a point on the base BC of the (Euclidean) triangle ABC such that the angle ADB is acute and the segment BD is shorter than the segment DC. Show that the side AB is shorter than the side AC.

   Hint. Join M, the midpoint of BC to A and use the inequality theorems of triangles listed in the introduction to the problems of Chapter 4. Alternatively, use the cosine rule.

   ii. P is a point inside a circle **C** and Q is its inverse with respect to **C**. **K** is a circle through P and Q, intersecting **C** at A and B. Show that the (Euclidean) length of the arc AQ is greater than that of the arc AP. (Note that the arcs AP and AQ are Poincaré congruent if the circles **C** and **K** give rise to Poincaré lines in **P**.)

   iii. $A_1, A_2, A_3, \ldots, A_n$ are distinct points on a Poincaré line ℓ such that $A_1 A_2, A_2 A_3, \ldots, A_{n-1} A_n$ are Poincaré congruent. Let P be any point on ℓ such that A_2 lies between A_1 and P. Show that, for n sufficiently large, P lies between A_1 and A_n. (This corresponds to the Archimedean Axiom—see the axioms in Section I of Part II.)

9. Let ℓ be a Poincaré line, its ideal points being A and B. Let Q be a Poincaré point not on ℓ. The Poincaré line m through Q, orthogonal to ℓ, intersects ℓ in L.

   i. Show how to construct a Poincaré line n through Q intersecting **P** at the ideal point A. (This is one of the two "critical" parallels through Q to ℓ.) The angle between the Poincaré segments QL and QA is the angle of parallelism at Q.

   ii. Show that the *angle of parallelism remains unchanged* if Q is replaced by a point P *on a diameter of* **P** orthogonal to ℓ and intersecting ℓ at C, P being chosen so that the Poincaré segments PC and QL are Poincaré congruent, the angle of parallelism being now the angle between the Poincaré segments PC and PA. This result means that the angle of parallelism depends only on the "Poincaré distance" of the external point Q from the line ℓ.

   iii. Let P move from C towards S, one of the two ideal points of the diameter through C. Show that for each position of P the angle of parallelism is an acute angle which decreases from $90°$ to $0°$ as P moves from C to S. (P is chosen so that C lies between O, the centre of **P** and P.) (Note that as the Euclidean distance CP increases so does the Poincaré distance of P from C along the line CS, though the Poincaré distance is *not* proportional to the Euclidean distance.)

Note. As already suggested in the conversations, "Poincaré distance" is identified with *inverse distance*. See Question 11.

10. *Definition*: A Poincaré circle about the Poincaré point C is the set of all Poincaré points such that the Poincaré segments joining each of these points to C are Poincaré congruent.

(This corresponds to the definition of a circle in a Euclidean plane.)

Consider the Poincaré lines going through a fixed Poincaré point C. Let $\overline{C}$ be the inverse of C in **P**. Extend the Poincaré lines through C into (Euclidean) circles. We obtain an elliptic family $\mathcal{E}$ of circles intersecting in C and $\overline{C}$. Consider those circles belonging to the hyperbolic family $\mathcal{H}$ associated with $\mathcal{E}$, i.e., the set of circles orthogonal to each circle in the family $\mathcal{E}$. Let **K** be a circle belonging to the family $\mathcal{H}$ and *inside* **P**. (Note that **P** belongs to $\mathcal{H}$, hence no circle of $\mathcal{H}$ intersects **P**.) Show *that* **K** *is a Poincaré circle of centre* C. Are there any Poincaré circles of different type?

11.

i. The *inversive distance* of two *concentric* circles of centre O and radii a and b is defined as
$$d = |\log a - \log b|,$$
(where the base of the logarithms is e).

   a. Show that d remains *unchanged* when the two circles are inverted in any circle about the common centre O.

   b. Show also that if a, b, c are the radii of three concentric circles such that $a < b < c$, or $a > b > c$, and d_1, d_2, d_3 denote the inversive distances between a and b, b and c, and a and c, respectively, then
$$d_3 = d_1 + d_2.$$

ii. **P** is a circle of centre O and radius r, $P_1 P_2$ and $E_1 E_2$ are two perpendicular diameters of **P**, and A is a point on the segment OE_2. **C** is a circle intersecting orthogonally both the diameter $E_1 E_2$ and the circle **P** in A and T, respectively. **K** is a circle of centre E_1 and passing through E_2.

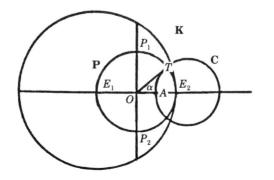

    a. Find the inverses in **K** of the lines $E_1 E_2$ and $P_1 P_2$, of the circles **P** and **C**, and of an arbitrary circle through E_1 and E_2.

    b. If a is the radius of **C**, and $\angle TOA = \alpha$, show that the radius of the inverse of **C** is $2r \tan \frac{\alpha}{2}$.

Hint. Express trigonometrical ratios in terms of $t = \tan \frac{\alpha}{2}$.

iii. Let A and B be Poincaré points on the segment OE_2. The inversive distances of A and B from O were defined as

$$(OA) = \log_e \cot \frac{\alpha}{2}$$

and similarly

$$(OB) = \log_e \cot \frac{\beta}{2}$$

(where β is defined similarly to α). Define the inversive distance (AB) as

$$|(OA) - (OB)| = \left| \log_e \cot \frac{\alpha}{2} - \log_e \cot \frac{\beta}{2} \right|.$$

    a. Show that if two segments AB and CD on OE_2 are Poincaré congruent then $(AB) = (CD)$.

    b. Show also that if B is a (Poincaré) point *between* A and C then

$$(AB) + (BC) = (AC).$$

Hint. Use (i) and (ii).

Chapter 6

Into the Shadows

THE time has come," friend Whatif said.
"Not again!" sighed Alice.
"To say goodbye," continued he,
"I'll leave your gentle valleys.

Your plains (or planes, Euclidean?),
Your winding rivers, meadows.
I'll leave your circle—family,
I'll vanish in the shadows."

Alice heard the voice, but it was too dark at first to see Dr. Whatif or to distinguish any shadows. Then a tiny pointlike light-source lit up somewhere behind Alice and she exclaimed, "Oh, the shadows! I can see Humpty Dumpty!"

"Humpty Dumpty?" the voice of Dr. Whatif came laughingly. "My dear young lady! What you can see there is your own shadow. It is a great pity that my time is up, I would have liked to take you for a nice long journey to give you a taste of what we call *projective geometry*."

Alice was not at all amused. She turned in the direction of the voice. "That shadow! Surely, it is not mine!"

The disembodied voice of Dr. Whatif now broke into wild laughter. "Hahaha. But it is! It is! It is a—a

comic section!"

The faint light spread now a little, and as the darkness lifted, the funny shadows waned and Alice perceived the figure of Lewis Carroll holding a large double cone in his hands.

"*Conic* sections are not new to us," said L.C., and with deft movements he sliced out plane sections from his cone and dropped the sections at Alice's feet.

"A circle!" shouted Alice,

"an ellipse!"

"another ellipse"

"a parabola"

as L.C. tilted his saw, and

"a hyperbola!"

as the saw turned more upright and the section fell into two disjointed pieces tumbling to the ground.

Dr. Whatif's voice came from the background. "You accept that all those different shapes come from the same cone. Instead of wood you could have your cone made out of rays of light coming from a point."

"A light cone?

"You will get a great variety of shadows, Alice and Humpty Dumpty, yet ...," Dr. Whatif stopped for a moment.

"Yet?"

"It is interesting to investigate the things that remain unchanged in the shadows."

"But the shadow is not true to size," objected Alice.

"True! I mean, it is not true," agreed Whatif.

"Not true to shape either."

"Admitted. So we will not be interested in shapes and sizes."

"What is left then?" Alice was puzzled.

This time Lewis Carroll smiled and held out a straight stick in front of his little torch. The shadow was longer than the stick, but still straight. Alice was not particularly impressed.

"May I borrow some knitting needles from you?" asked L.C.

Alice started to say, "I do not carry them with me," but before she could utter the sentence, there were the needles, forming a neat star in front of the small light casting a star-like shadow on the wall. "The shadows of the needles still *pass* through the same point," observed Alice.

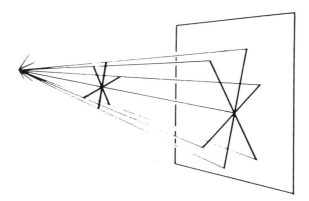

From the background came applause. It must have come from Dr. Whatif, still unseen.

"That will be precisely our concern now. You must learn to see things in *perspective*." He continued, "The point source of the light, or the vertex of that cone are centres of *perspectivity*. Through perspectivity many things can change: a circle may turn into an ellipse, or even a hyperbola, Alice may turn into Humpty Dumpty, but," and here his voice turned emphatic

"1. If a number of points lie on a line, their images also lie on a line.

2. If a number of lines pass through the same point, their images also pass through a point."

"But the objects become so, so transformed!" objected Alice.

"Sure, we are dealing with *transformations*, don't forget. Just recently we discussed congruence. A configuration is congruent to another one if a finite number of reflections transforms the first configuration into the second one. The transformations you are observing now are perspectivities."

Light flooded the place and Alice could clearly see Dr. Whatif who made sure that she could also clearly see what a perspectivity is. He sketched the figure below.

"You can see the two planes $\mathcal{P}_1$ and $\mathcal{P}_2$ meeting in the line a that we call the *axis of perspectivity*. The point V is the *centre of perspectivity*. Take a point, say X_2, of the configuration f_2 in the plane $\mathcal{P}_2$. It is the perspective image of a point X_1 of the configuration f_1 in the plane $\mathcal{P}_1$. It simply means that the line VX_1 hits

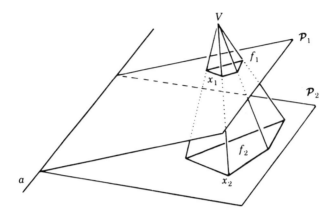

the plane $\mathcal{P}_2$ in point X_2. In this way we *project* all points of f_1 onto the plane $\mathcal{P}_2$, obtaining f_2 as the perspective image of f_1 and, of course, a, the axis, is projected into itself."

Alice asked, "Can we project *all* the points of $\mathcal{P}_1$ to the plane $\mathcal{P}_2$?"

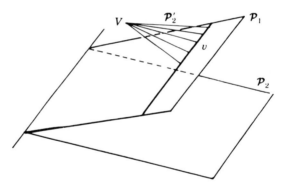

Dr. Whatif went on unperturbed. "We can place a plane $\mathcal{P}_2'$ through V which, by Euclidean geometry, is parallel to $\mathcal{P}_2$. So if v is the line in which $\mathcal{P}_2'$ meets the plane $\mathcal{P}_1$, then, again by Euclidean geometry, no point of v has a perspective image on $\mathcal{P}_2$. If V is a light source, then the image of v on $\mathcal{P}_1$ ...," he halted and let Alice finish the sentence.

"vanishes."

"So?"

"We could invite infinity again, couldn't we?"

"That is precisely what we do. We *extend* all the Euclidean planes to infinity," answered Whatif.

"Just like we did when we needed the inverse of the centre of the circle of inversion," said Alice brightly.

"Well, not quite. In the inversive plane we needed just one point for our infinity.

Here we have a whole *ideal line*, our line of infinity."

This time Alice was quick to catch on. "Any line in the plane will hit the ideal line somewhere, after all we can go on and on along that line, and . . . ."

"And?"

"And if two lines of a plane meet the ideal line in the same point, then they are parallel."

"Good! What made you think so?" L.C. interjected. He was proud of Alice.

"They could not be intersecting lines, intersecting in an ordinary point, because then we would have two lines intersecting in two different points!"

"So we have arrived to the *projective plane*, where things are very simple indeed," concluded Whatif and added

"1. Through each pair of points there is exactly one line.

2. Each pair of lines intersects in one point."

"So we have no vanishing lines?"

"Conveniently, we still call that line v in our diagram the vanishing line of that perspectivity, but you and I will know that its image in $\mathcal{P}_2$ is the *ideal line* of $\mathcal{P}_2$."

Alice brightened. "And I suppose, we can reverse this. The ideal line of $\mathcal{P}_2$ could be *projected* into $\mathcal{P}_1$ to obtain v, an ordinary line of $\mathcal{P}_1$, and then all those parallel lines of $\mathcal{P}_2$, meeting on the ideal line, have images in $\mathcal{P}_1$ that actually meet!"

"You get the idea," beamed Dr. Whatif, "I have no more time left, but perhaps one day someone else may take you on a journey into projective geometry, or even to some other geometries we have developed in the twentieth century."

At this the lights went out suddenly, whimsical light-spots cast weird shadows. Suddenly Alice recognised a familiar shape. A pleading voice came from its direction.

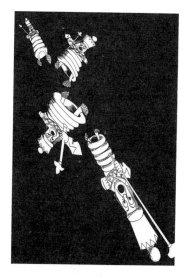

"Help, Alice, help. Help an old friend! I know that on you I can always depend."

"It must be the White King," thought Alice, "the poor thing always gets into trouble."

The figure seemed to tumble in the flashing lights, his shape changing wildly.

"Help, Alice, help."

"Oh, what distress!"

"My poor crown, my sceptre!"

"Oh, what a mess."

"Help, Alice, help."

"He is really almost beyond recognition," Alice thought compassionately. "If I only could restore him by a suitable projection!"

Was Dr. Whatif or Lewis Carroll pushing her hand or did she suddenly become so clever?

She found somehow the right centre of perspectivity V and a kind plane $\mathcal{P}$, and lo and behold, there was the White King in his full glory, original size, and shape.

The figure smiled and faded out peacefully in the full light. Dr. Whatif was standing next to Alice.

He had encouraging words for her. "Bravo! You recovered his sceptre from infinity. Of course, you could only do it because his projective features did not change. If you journey into projective geometry, life becomes very simple indeed. In a projective plane we have our points and lines."

"What are they actually?"

"You should know by now that we do not worry about defining those! We only need *three axioms* on which to base the geometry of the *projective plane*. You have already had two of them, but I repeat them together with the third one:

1. Through any two distinct points there is exactly one line.
2. Any two distinct lines intersect in a point.
3. There exist four points such that no three of them lie on a line."

This brought back to Alice the memory of her first meeting with Whatif. "Those axioms can be satisfied by a set of just a few points and lines, like in the 'gemmetry' of the seven stones and seven chains."

"I am glad you remember. Projective geometry is so versatile that it can describe your Euclidean plane, extended by the ideal line, and so illuminate the connection between things that seem to be so different on first sight as, for example, the circle and the hyperbola. Alternatively it can be applied to a situation involving a *finite* number of objects such as the stones and chains in our 'gemmetry,' arranged by some design."

Alice looked thoughtful. She tried to remember the details of that 'gemmetry.' Whatif guessed her thoughts.

"You can forget now your diamonds and amethysts. Instead, we shall speak about you, Alice, and your young friends: Beatrice, Catherine, Dorothy, Emily, Flora, and Geraldine. Suppose that each of you wants to meet each friend every week in that nice little tea-shop where they can seat just three of you at the little tables. So you arrange different meeting days for three girls at a time. You may have a schedule like the following:

Monday: Alice, Beatrice, Dorothy
Tuesday: Beatrice, Catherine, Emily
Wednesday: Catherine, Dorothy, Flora

Thursday: Dorothy, Emily, Geraldine
Friday: Emily, Flora, Alice
Saturday: Flora, Geraldine, Beatrice
Sunday: Geraldine, Alice, Catherine."

"What are the points and lines?"

"You see how good it is that we did not use restrictive definitions? But we could illustrate the situation with a diagram using 7 points A, B, C, D, E, F, G and 7 lines m, tu, w, th, f, sa, su. You can check that all three axioms of the projective plane are satisfied!"

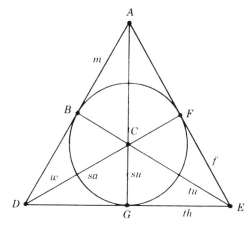

"Could you achieve this sort of thing with more than seven girls?" asked Alice.

"It will be harder to draw a diagram, but you could have a set of 13 friends, each of whom meets every other friend once in 13 days and if you pick any two of the 13 days, there will be exactly one girl who goes to a meeting on each of the two days.

Moreover you can find four girls such that no three of them meet on the same day. This time there are four girls who meet each day."

"How about other numbers?"

"The next number is 21, and we can easily construct infinitely many others. You can do it yourself with some little hints. I shall leave a small problem package as a farewell present."

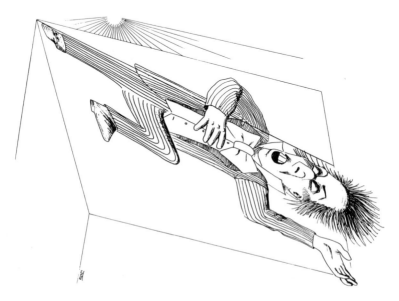

Those lights! They went out again and Alice saw in the dim light the plane $\mathcal{P}_1$ with Dr. Whatif stretched along line v. The light was now thrown onto plane $\mathcal{P}_2$, but Dr. Whatif could not be seen any more. Alice rubbed her eyes. It was brilliant daylight. There was no Dr. Whatif and she could not see Lewis Carroll either. Was it a dream? Was it a weird but real journey into those geometries? Alice thought it was a dream when she noticed a small problem package at her feet.

"It must have been a journey," she thought dreamily, "later I may even have similar journeys, but first I would like to travel to Paris."

Problems and Exercises 6

Preliminary Remarks:

(a) *The extended Euclidean space.* As discussed in the last chapter, every line is endowed with a unique *ideal point*, the "point at infinity." Each family of *parallel* lines has a *common ideal point, distinct* from the ideal points belonging to other parallel families. The set of ideal points of a plane is the *ideal line* of the plane. Each ordinary line of the plane intersects the ideal line in one point (namely the ideal point of the ordinary line), hence the ideal line has no ordinary points. The ideal line of a plane is common to all planes parallel to that plane and distinct from the ideal lines of planes not parallel to it. Finally, the set of ideal lines is the ideal plane of the space.

When referring to the extended Euclidean space, we replace the word "parallel" by "intersecting in the ideal point," when speaking about lines, or "intersecting in the ideal line," when dealing with parallel planes. It is not difficult to check that this change of language does not affect the properties established in Euclidean geometry. Theorems concerning for example, parallelograms or similar triangles or other configurations involving parallel lines still remain valid. Moreover, such theorems can be used advantageously for more general configurations when dealing with properties involving intersections of lines or collinearity of points which remain unaltered by perspectivities.

(b) *Projective transformations.* In Chapter 5 congruence transformations were discussed in some detail. A transformation assigns to every point of the plane, or more generally of the space, an image point in that given plane, or more generally in space. Congruence in the plane was defined as a transformation brought about by a succession of reflections in lines. A *projective transformation* in the *extended Euclidean* space is a transformation brought about by a succession of perspectivities. In the text, the centre of perspectivity V was an ordinary point. Instead of a point source of light, we may have parallel rays and consider shadows cast by these parallel rays. We say then that the centre of perspectivity is the ideal point at which the parallel rays meet and call the perspectivity a *parallel perspectivity.* Like perspectivities with centres at ordinary points, parallel perspectivities transform configurations in such a way that both collinearity of points and concurrence of lines are preserved. In either case images of points may be ordinary or ideal points and images of ideal points can be ordinary points.

Definition. The geometrical configurations A and B are *projectively equivalent* if we can arrive from A to B through a succession of perspectivities.

1.

a. Show that if two plane figures A and B are congruent, then they are also projectively equivalent. (*Hint.* Use the definition of congruence of Chapter 5.)

b. Show that two similar triangles are projectively equivalent.

2. $ABCD$ is a quadrangle in plane $\mathcal{P}_1$. Show that by a suitable choice of the centre of perspectivity V, the image of $ABCD$ in some plane $\mathcal{P}_2$ is a parallelogram.

In this problem and all the others in this section, the quadrangles in question are nondegenerate, that is, no three of the four vertices are on the same line.

3.

a. Find a perspectivity such that the image of a given parallelogram is a square.

b. Show that all quadrangles are projectively equivalent to a fixed square and hence all quadrangles are projectively equivalent.

4. Let ABC be a triangle in the plane $\mathcal{P}_1$ and $A'B'C'$ its perspective image in plane $\mathcal{P}_2$, the centre of perspectivity being V (a point not in $\mathcal{P}_1$ or $\mathcal{P}_2$). Show that if the intersections of the line-pairs BC and $B'C'$, AC and $A'C'$, AB and $A'B'$ are X, Y, Z, respectively, then X, Y, Z are collinear. Conversely, if X, Y, Z are known to lie on the same line, then the triangles ABC and $A'B'C'$ are in perspective.

This is Desargues' theorem for the case when the triangles ABC and $A'B'C'$ are not in the same plane.

5. Show that Desargues' theorem is also valid in the case when the triangles ABC and $A'B'C'$ are in the same plane.

Hint. Step 1. Assume first that X, Y, Z are collinear (using notations of problem 4). Let a be the line XYZ. Let $\mathcal{P}_1$ be a plane through a distinct from $\mathcal{P}$ (the plane of triangles ABC and $A'B'C'$). Let $A_1B_1C_1$ be the perspective image of ABC in $\mathcal{P}_1$, the centre of perspectivity being V, a point not in $\mathcal{P}$ or $\mathcal{P}_1$. Show that $A_1B_1C_1$ and $A'B'C'$ are in perspective, the centre being some point V'.

Step 2. Show that the line VV' intersects $\mathcal{P}$ in the centre of perspectivity of triangles ABC and $A'B'C'$.

Step 3. Having proved one part of the theorem, prove the converse, i.e., that two triangles, known to be in perspective from a *point*, have corresponding sides intersecting at points on a line.

6. (This is an easier way to prove Desargues' theorem for coplanar triangles.) We use the same notations as in Problems 4 and 5. Prove Desargues' theorem by using the fact that the quadrilateral $BCC'B'$ is projectively equivalent to a parallelogram. (You may assume $BCC'B'$ is nondegenerate for if, e.g., C and C' coincide then the theorem is trivially true.)

7. Pappus's Theorem. A_1, B_1, C_1 are three distinct points on line ℓ_1, A_2, B_2, C_2 are three distinct points on line ℓ_2, and the intersections of the line-pairs B_1C_2 and C_1B_2, A_1C_2 and A_2C_1, A_1B_2 and A_2B_1 are A_3, B_3, C_3, respectively. Show that A_3, B_3, C_3 are on the same line ℓ_3.

Hint. First show that if $A_1B_2 \| A_2B_1$ and $A_1C_2 \| A_2C_1$, then $B_1C_2 \| B_2C_1$. Distinguish between two cases:

   i. lines A_1, B_1, C_1 and A_2, B_2, C_2 intersect;

   ii. they are parallel.

Next use projective transformation to establish the general theorem.

8. A set of a finite number of points and lines satisfies the three axioms of the projective plane as stated in the text.

Suppose that the set contains a line which contains exactly k points. Prove that

   i. every line of this projective plane contains exactly k points;

   ii. exactly k lines intersect in each point;

   iii. the total number of points is $k^2 - k + 1$;

   iv. the total number of lines is $k^2 - k + 1$.

Note. As in the example described in the text, the points may represent members of a certain club, the lines their meeting days. For reasons given below, the number describing the finite projective plane is denoted by q, where $k = q + 1$, so that the number of points on each line is $q + 1$, the number of lines through each point $q + 1$ and the total number of points, also the total number of lines of the plane is given by $q^2 + q + 1$.

Projective planes are known to exist for all those values of q which are powers of a prime number, hence for

$$q = 2, \ 3, \ 4, \ 5, \ 7, \ 8, \ 9, \ 11, \ 13, \ 16, \ 17, \ 19, \ 23, \ 25, \quad \text{and so on.}$$

No other finite projective planes are known so far.

Part II

Solutions
to the
Problems and Exercises

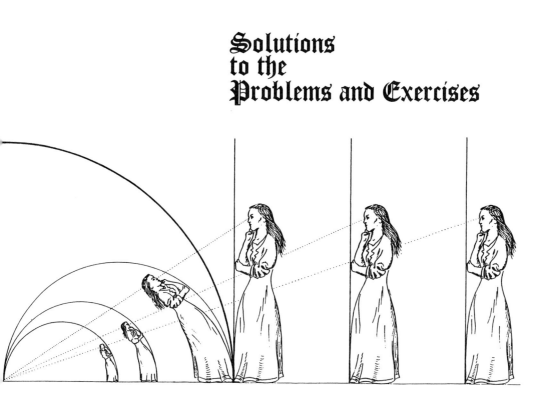

Section I

Axiom Systems

1. Introduction

The first part of this book gives an informal, discursive treatment of the logical basis of geometry, or more appropriately "geometries," since the nineteenth and twentieth centuries produced alternatives to classical Euclidean geometry. Since Euclid reigned supreme for two thousand years, it is of importance to say at least a few words about his *Elements* the work on which the whole structure of our geometry has been resting almost to this day. The few cracks which have been discovered in its foundation, have been sufficiently mended to ensure that the building still stands solid and stable.

Euclid's *Elements* summarises and systematises Greek contributions to mathematics. Great Greek geometers such as Thales, Pythagoras and others turned the hitherto empirical science of geometry into a deductive logical system. The work, *Elements,* appeared around 300 B.C. It consists of thirteen books, contains 465 propositions, dealing with geometry and arithmetic, which with the addition of elementary algebra (the Hindu-Arab contribution) comprise the major part of the body of mathematical knowledge transmitted in general education up to our days. The first book of the *Elements* contains the definitions and basic axioms (or postulates). In modern geometry most of these definitions are discarded and replaced by the notion of a few "primitive elements" which are left undefined and serve as a basis for defining other concepts. It was also necessary to complete the axiom system of Euclid by axioms of order ("betweenness") and continuity, without which there have been gaps in the proofs of geometrical theorems.

Only the five basic postulates of Euclid will be quoted here as relevant for this book. The first four of these were accepted as postulates throughout the centuries leading to the nineteenth century, the time of Gauss, Bolyai, Lobachevsky, but the fifth postulate caused controversy. It was felt that it is a *theorem*, which should be deduced from more basic definitions and axioms and, in fact, many "proofs" were given, all of them false. The advent of hyperbolic geometry, based on Euclid's postulates—with the exception of the fifth which it replaced by an axiom *contradicting* the Euclidean axiom and yet resulting in a geometrical system as consistent as the Euclidean one—showed that the fifth postulate is indeed *independent* of the others.

"Absolute geometry" is the system which leaves the *question of parallelism open*, and is based on the remaining axioms. As pointed out in the problem section of Chapter 4, there is a great number of theorems which are equally valid in both the Euclidean and hyperbolic geometry.

Chapter 6 of this book gives a glimpse of *projective geometry*, a system which is based on very few primitive elements and axioms, doing without notions of parallelism, order, and continuity. Yet it gives rise to a rich collection of theorems and illuminating relationships between structures which are also subjects of Euclidean geometry, for example, the conic sections: circle, ellipse, hyperbola, parabola. Projective geometry is another exciting development of modern geometry. However, it is beyond the scope of this book and is touched on only as an example of alternative geometrical structures. In the following a short excerpt from the *Elements* (the five postulates) and the Euclidean axiom-system of Hilbert will be given. Alternative modern axiom-systems, equivalent to Hilbert's, can also be found in the literature.

THE POSTULATES OF EUCLID

Let the following be postulated:
1. A straight line can be drawn from any point to any point.
2. A finite straight line can be produced continuously in a straight line.
3. A circle may be described with any centre and distance.
4. All right angles are equal to one another.
5. If a straight line falling on two straight lines makes the interior angles on the same side together less than two right angles, the two straight lines, if produced indefinitely, meet on that side on which the angles are together less than two right angles.

HILBERT'S AXIOMS FOR PLANE EUCLIDEAN GEOMETRY

Primitive Terms: points, line, on, between, congruent.

GROUP 1: AXIOMS OF CONNECTION
1.1   There is one and only one line passing through any two given distinct points.
1.2   Every line contains at least two distinct points, and for any given line there is at least one point not on the line.

GROUP 2: AXIOMS OF ORDER

2.1   If point C is between points A and B, then A, B, C are all on the same line, and C is between B and A, and B is not between C and A, and A is not between C and B.

2.2   For any two distinct points A and B there is always a point C which is between A and B, and a point D such that B is between A and D.

2.3   If A, B, C are any three distinct points on the same line, then one of the points is between the other two.

Definitions. By the segment AB is meant the points A and B and all points which are between A and B. Points A and B are called the *end points* of the segment. A point C is said to be *on* the segment AB if it is A or B or some point between A and B.

Definition. Two lines, a line and a segment, or two segments, are said to *intersect* if there is a point which is on both of them.

Definitions. Let A, B, C be three points not on the same line. Then by the *triangle ABC* is meant the three segments AB, BC, CA. The segments AB, BC, CA are called the *sides* of the triangle, and the points A, B, C are called the *vertices* of the triangle.

2.4   (Pasch's axiom). A line which intersects one side of a triangle but does not pass through any of the vertices of the triangle must also intersect another side of the triangle.

GROUP 3: AXIOMS OF CONGRUENCE

3.1   If A and B are distinct points and if A' is a point on a line m, then there are two and only two distinct points B' and B'' on m such that the pair of points A', B' is congruent to the pair A, B and the pair of points A', B'' is congruent to the pair A, B; moreover, A' is between B' and B''.

3.2   If two pairs of points are congruent to the same pair of points, then they are congruent to each other.

3.3   If point C is between points A and B and point C' is between points A' and B', and if the pair of points A, C is congruent to the pair A', C', and the pair of points C, B is congruent to the pair C', B', then the pair of points A, B is congruent to the pair A', B'.

Definition. Two segments are said to be *congruent* if the end points of the segments are congruent pairs of points.

Definitions. By the *ray AB* is meant the set of all points consisting of those which are between A and B, the point B itself, and all points C such that B is between A and C. The ray AB is said to *emanate from* point A.

Theorem. *If B' is any point on the ray AB, then the rays AB' and AB are identical.*

Definitions. By an *angle* is meant a point (called the *vertex* of the angle) and two rays (called the *sides* of the angle) emanating from the point. By virtue of the above theorem, if the vertex of the angle is point A and if B and C are any two points other than A on the two sides of the angle, we may unambiguously speak of the angle BAC (or CAB).

Definitions. If ABC is a triangle, then the three angles BAC, CBA, ACB are called the *angles* of the triangle. Angle BAC is said to be included by the sides AB and AC of the triangle.

3.4 If BAC is an angle whose sides do not lie on the same line, and if A' and B' are two distinct points, then there are two and only two distinct rays, $A'C'$ and $A'C''$, such that angle $B'A'C'$ is congruent to angle BAC and angle $B'A'C''$ is congruent to angle BAC; moreover, if D' is any point on the ray $A'C'$ and D'' is any point on the ray $A'C''$, then the segment $D'D''$ intersects the line determined by A' and B'.

3.5 Every angle is congruent to itself.

3.6 If two sides and the included angle of one triangle are congruent, respectively, to two sides and the included angle of another triangle, then each of the remaining angles of the first triangle is congruent to the corresponding angle of the second triangle.

GROUP 4: AXIOM OF PARALLELS

4.1 (Playfair's axiom). Through a given point A not on a given line m there passes at most one line which does not intersect m. (Note: This is equivalent to the fifth postulate of Euclid.)

GROUP 5: AXIOMS OF CONTINUITY

5.1 (Axiom of Archimedes). If A, B, C, D are four distinct points, then there is, on the ray AB, a finite set of distinct points, $A_1, A_2, \ldots, A_n$ such that
(1) each of the pairs $A, A_1; A_1, A_2; \ldots; A_{n-1}, A_n$ is congruent to the pair C, D, and
(2) B is between A and A_n.

5.2 (Axiom of Completeness). The points of a line constitute a system of points such that no new points can be assigned to the line without causing the line to violate at least one of the nine axioms 1.1, 1.2, 2.1, 2.2, 2.3, 2.4, 3.1, 3.2, 5.1.

Solutions to the Problems

Chapter 1

1. If P is on the circle, then $d = r$. So $P(\mathbf{C}) = 0$. If $A_1 A_2$ is any chord through P, then $P = A_1$ or $P = A_2$, hence $PA_1 \cdot PA_2 = 0 = P(\mathbf{C})$. If P is inside the circle and $A_1 A_2$ a chord through P, then draw the diameter $B_1 B_2$ through P.

$\triangle P A_1 B_2 \sim \triangle P B_1 A_2$ (equiangular), so

$$PA_1/PB_1 = PB_2/PA_2$$

or

$$PA_1 \cdot PA_2 = PB_1 \cdot PB_2.$$

Here $PB_1 = r - d$ and $PB_2 = r + d$, hence

$$PA_1 \cdot PA_2 = (r - d)(r + d) = r^2 - d^2 = |P(\mathbf{C})|.$$

2. Let $\mathbf{C}_1$ and $\mathbf{C}_2$ be the given circles and $\mathbf{C}$ intersecting both orthogonally.

Let S be the centre of $\mathbf{C}$ and T_1, T_2 intersection points of $\mathbf{C}$ and $\mathbf{C}_1$ and $\mathbf{C}$ and $\mathbf{C}_2$, respectively. Since $\mathbf{C}$ is orthogonal to $\mathbf{C}_1$ and $\mathbf{C}_2$, the tangents to $\mathbf{C}_1$ and $\mathbf{C}_2$ at T_1 and T_2, respectively, pass through S. Thus $ST_1 = ST_2$ and so S is on the radical axis of $\mathbf{C}_1$ and $\mathbf{C}_2$.

Conversely, if S is on the radical axis and $\mathbf{C}$ is known to intersect $\mathbf{C}_1$ orthogonally, the line ST_1 is a tangent to the circle $\mathbf{C}_1$. Since S is on the radical axis of $\mathbf{C}_1$ and $\mathbf{C}_2$, then $ST_2 = ST_1$, where ST_2 is the tangent from S to $\mathbf{C}_2$. It follows that T_2 is on the circle $\mathbf{C}$ so $\mathbf{C}$ intersects $\mathbf{C}_2$ orthogonally.

3. Let $O_1A = x_1$ and $O_2A = x_2$. (See Figure 1.) Then $x_1^2 - r_1^2 = x_2^2 - r_2^2$, so $x_1^2 - x_2^2 = (x_1 + x_2)(x_1 - x_2) = s(x_1 - x_2) = r_1^2 - r_2^2$.

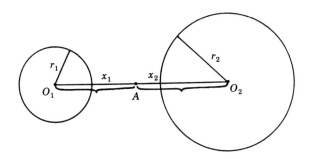

Figure 1.

Thus $x_1 - x_2 = \dfrac{r_1^2 - r_2^2}{s}$ and $x_1 + x_2 = s$. Solving the equation, obtain $x_1 = \dfrac{s^2 + r_1^2 - r_2^2}{2s}$ and $x_2 = s - x_1 = \dfrac{s^2 + r_2^2 - r_1^2}{2s}$. Assume that $\mathbf{C}_2$ lies inside $\mathbf{C}_1$. The point A lies outside $\mathbf{C}_1$. Denote $O_1A = x_1$, $O_2A = x_2$. Using notations as above, we have $s = x_1 - x_2$. From $x_1^2 - r_1^2 = x_2^2 - r_2^2$ as above, we have now $s(x_1 + x_2) = r_1^2 - r_2^2$, resulting in

$$x_1 = \frac{r_1^2 - r_2^2 + s^2}{2s}$$

and

$$x_2 = \frac{r_1^2 - r_2^2 - s^2}{2s}.$$

4.

i. Since the centre of any circle of the given elliptic family is equidistant from A and B, it lies on the line bisecting AB perpendicularly. Hence all the centres lie on the same line. (See Figure 2.)

ii. Let P be the intersection of ℓ_1 and ℓ_2. Then

$$P(\mathbf{C}_2) = P(\mathbf{C}_3), \quad \text{and} \quad P(\mathbf{C}_3) = P(\mathbf{C}_1).$$

It follows that

$$P(\mathbf{C}_1) = P(\mathbf{C}_2),$$

hence P is on ℓ_3.

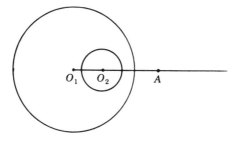

Figure 2.

a. Since the radical axis of two circles is perpendicular to the line joining the centres, it follows that ℓ_1, ℓ_2, and ℓ_3 are parallel.

b. As the distance of the centres of two circles approaches zero, it follows from the formulae in Problem 3 that the radical axis moves to infinity.

(In both cases (a) and (b) we may say that the three radical axes meet in infinity.)

5. Let C_1 and C_2 be two nonintersecting circles. Construct any circle C_3 intersecting both C_1 and C_2. (This is always possible.) The common chords of C_1 and C_3 and of C_2 and C_3, respectively, intersect at some point P. Then P lies on the radical axis of C_1 and C_2, so the radical axis is obtained by drawing through P a line perpendicular to the join of the centres of C_1 and C_2.

Alternatively: construct another circle C_3' intersecting C_1 and C_2 and obtain the point P' in the same manner as P. The radical axis is the line PP'.

6. Each circle of the family $\mathcal{E}$ has AB as a chord, hence each circle in $\mathcal{E}$ is of diameter greater than or equal to $AB = a$. Thus the smallest circle of $\mathcal{E}$ is the one with diameter AB. Denote this circle by C_0.

Each circle in $\mathcal{H}$ intersects C_0 orthogonally. If d is the distance between the centre of one circle of $\mathcal{H}$ and the centre of AB, then

$$d^2 - r^2 = \left(\frac{a}{2}\right)^2,$$

where r is the radius of the circle belonging to $\mathcal{H}$. Hence, $d > \frac{a}{2}$.

When $d = \frac{a}{2}$, then $r = 0$. So in this case the smallest circle is a point, coinciding with A or B.

The radius of a circle in $\mathcal{E}$ or $\mathcal{H}$ can be of any length greater than the minimal value established above. As the radius (in $\mathcal{E}$ or $\mathcal{H}$) goes to infinity, the circle tends to the radical axis of the family.

7. Let OT be tangent from O to any of the circles in $\mathcal{E}$. (See Figure 3.) Then

$$OT^2 = OA \cdot OB.$$

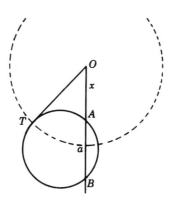

Figure 3.

Here $OT = r$, $OA = x$, $OB = x + a$. Hence

$$r^2 = x(x + a)$$

or

$$ax + x^2 = r^2,$$

whence $ax \leq r^2$, the equality holding only in the case when $x = 0$ (or O coinciding with A).

If r is kept constant while a increases to infinity, x tends to O. Hence circles of radius r, belonging to the (changing) family $\mathcal{H}$, tend to the circle about A and of radius r, while the family $\mathcal{E}$ tends to a family of lines through A. (See diagram near the end of the chapter.)

8. The centre S of the required circle is on t, the line through A, tangent to **C**. It is also on m, the line perpendicularly bisecting AP, hence it is the intersection of t and m.

This intersection determines uniquely the required circle.

The construction fails if ℓ and m are parallel, that is, if PA is pependicular to t. That happens if the centre of **C** lies on AP. In that case, the line AP can be regarded as a circle of infinite radius orthogonal to **C**.

9. Let $\mathbf{C}_1$ be any of the circles through P, orthogonal to **C**, and intersecting **C** at T. Let OP also intersect $\mathbf{C}_1$ in P'. Since OT is a tangent to $\mathbf{C}_1$,

$$O(\mathbf{C}_1) = OT^2 = OP \cdot OP' \qquad (*)$$

Here $OT = r$ and OP are of fixed length, hence OP' is also fixed and therefore P' is the same for all the circles through P and orthogonal to **C**.

If P is very close to O, then it follows from (*) that OP' is large. If P coincides with O, it follows from (*) that OP' has no finite value. The orthogonal circles become lines (circles of infinite radii) intersecting in O only. (Having their second intersection points in infinity.)

10. On the line OA find the point A' such that

$$OA \cdot OA' = r^2,$$

where r is the radius of **C**.

(A' can be constructed by using the construction in Problem 8, choosing arbitrarily a point P on **C** such that P is not on AO. Alternatively, let the perpendicular to OA at A intersect **C** at the point T. Then the tangent line to **C**, touching **C** at T, intersects OA produced in A'.)

The required circle goes through the points A, A', and B, hence its centre is on the lines perpendicularly bisecting AA' and AB.

Note. If A, B and O are on the same line, the construction fails. In this case the diameter OAB can be regarded as a circle of infinite radius orthogonal to **C**.

11. The centres of all circles orthogonal to $\mathbf{C}_1$ and $\mathbf{C}_2$ are on the radical axis of $\mathbf{C}_1$ and $\mathbf{C}_2$ (see Problem 2), and circles orthogonal to $\mathbf{C}_2$ and $\mathbf{C}_3$ have their centres on the radical axis of $\mathbf{C}_2$ and $\mathbf{C}_3$. Hence the centre of the required circle is the intersection of these two radical axes. The solution is unique. The fact that the centres are not on the same line ensures that the radical axes are not parallel, hence the construction is possible.

12. Let O_1 and O_2 be the centres of $\mathbf{C}_1$ and $\mathbf{C}_2$, respectively. Let r_1 and r_2 be the radii of $\mathbf{C}_1$ and $\mathbf{C}_2$. On O_1P find the point P_1 such that $O_1P \cdot O_1P_1 = r_1^2$, and on O_2P find P_2 such that $O_2P \cdot O_2P_2 = r_2^2$ (see Problem 9). The required circle goes through P, P_1, and P_2 hence is determined uniquely, provided that P, P_1 and P_2 are not on the line O_1O_2. If P is on the line O_1O_2 (in particular, if O_1 and O_2 coincide), then the line O_1O_2P can be regarded as a circle of infinite radius intersecting $\mathbf{C}_1$ and $\mathbf{C}_2$ orthogonally.

Chapter 2

1. $\triangle SAP' \sim \triangle SPA$ (Common angle at S, right angles at A and P, respectively.) Hence $SP : SA = SA : SP'$, so $SP \cdot SP' = SA^2$ as claimed. Finding the inverse is done by reconstructing Figure 4, beginning with P in (a), and with P' in (b).

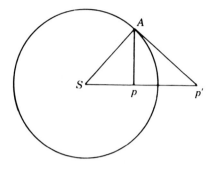

Figure 4.

2.

a. Let O_1 and O_2 be the centres of $\mathbf{C}_1$ and $\mathbf{C}_2$. Since $\mathbf{C}_1$ and $\mathbf{C}_2$ touch at S, the line O_1O_2 passes through S. The inverses of $\mathbf{C}_1$ and $\mathbf{C}_2$ are both lines perpendicular to SO_1O_2, hence they are *parallel lines*. (For finding their actual positions, it suffices to invert *one* point on each circle.)

b. Conversely, the inverses of two parallel lines are two circles touching at S and having their centres on the line through S perpendicular to the given parallel lines.

3. Referring to the figure in the text, let a and b be the two given lines, intersecting at P. Their inverses are two circles a' and b' intersecting at S, the centre of the circle $\mathbf{K}$, and at P', the inverse of P.

Denote the feet of the perpendiculars from S to a and b by A and B, respectively. Then SA and SB go through the centres of a' and b'.

The angle between a and b is denoted by $(ab) = \alpha$.

The angle between a' and b' is given by $(t_{a'}t_{b'}) = \alpha'$ where $t_{a'}$ and $t_{b'}$ are tangent lines at S to a' and b', respectively.

Since $t_{a'} \perp SA$, it follows that $t_{a'} \| a$. Similarly $t_{b'} \| b$. Hence $\alpha' = \alpha$.

Denote the tangent lines to a' and b' at P' (the inverse of P) by $\tau_{a'}$ and $\tau_{b'}$, respectively. Then, by symmetry about the line joining the centres of a' and b', the angle

$$(\tau_{a'}, \tau_{b'}) = (t_{a'}, t_{b'}) = \alpha' = \alpha.$$

Hence the magnitude of the angle between two lines remains unchanged by inversion.

(Note however that the sense of rotation from $\tau_{a'}$ to $\tau_{b'}$ is opposite to that from $t_{a'}$ to $t_{b'}$, hence from a to b.)

If $\mathbf{C}_a, \mathbf{C}_b$ are two circles intersecting at P with tangent lines a, b, their inverses $\mathbf{C}_{a'}, \mathbf{C}_{b'}$ intersect at P' (with $\tau_{a'}, \tau_{b'}$ being tangent lines to $\mathbf{C}_{a'}, \mathbf{C}_{b'}$ since tangent lines invert to tangent lines). Hence inversion does not change the angle between C_a and C_b.

The inverses of two tangential circles are again tangential circles, provided that neither of the circles passes through the centre of the circle **K** (circle of inversion).

If the point of contact of the two circles is the centre of **K**, then both circles invert into straight (parallel) lines. If only one of the circles goes through the centre of **K**, then its inverted image is a straight line tangential to the inverted image of the other circle.

4.   In Figures 5–8 below, the solid lines represent the figures to be inverted and the dotted lines show the inverses.

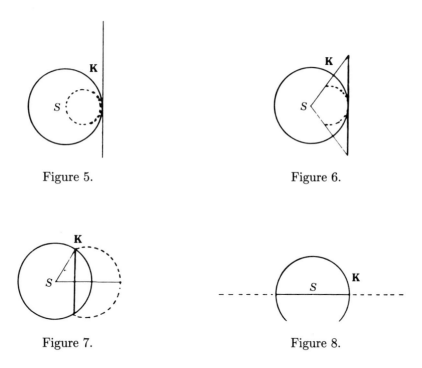

Figure 5.   Figure 6.

Figure 7.   Figure 8.

5.   Let PQ be one side of the polygon, touching **K** at T, and subtending angle α at S, the centre of **K**. (See Figure 9.) The inverse of the segment PQ is the arc of the circle of diameter ST. Bounded by SP and SQ, the arc subtends angle α at S. Thus the arc $P'Q'$ subtends an angle 2α at the centre of a circle of radius $k/2$, where k is the radius of **K**. So the arc $P'Q'$ is of length $k\alpha$, the *same as the length of the arc* of the circle **K** bounded by SP and SQ. Hence the length of the perimeter of the inverse of the polygon is equal to the circumference of **K**.

6.   Since P' is the inverse of P with respect to **K**, it follows that $(k-x)(k+y) = k^2$ hence

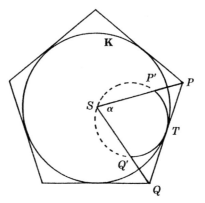

Figure 9.

$$\left(1 - \frac{x}{k}\right)\left(1 + \frac{y}{k}\right) = 1. \tag{1}$$

Let $r = \frac{y}{x}$. Writing $p = \frac{x}{k}$, equation (1) becomes

$$(1 - p)(1 + rp) = 1$$

whence $r = \frac{1}{1-p}$. Thus if $p \to 0$, $\frac{y}{x} = r \to 1$.

7. Let S be the centre of $\mathbf{K}$ and O the centre of $\mathbf{C}$. Let S be the origin of a rectangular coordinate system and SO the x-axis. Let A and B be the endpoints of the diameter of $\mathbf{C}$ along the x-axis. Denote the x-coordinates of A and B by a and b, respectively. Then the x-coordinate of O is

$$\frac{1}{2}(a + b).$$

We may assume here that a, b are different from 0, for otherwise the inverse of $\mathbf{C}$ is not a circle.

The inverses of A, B, O are on the x-axis. Denoting the x-coordinates of the inverses A', B', O' by a', b', σ', respectively, we have

$$a' = \frac{k^2}{a} \qquad b' = \frac{k^2}{b} \qquad \sigma' = \frac{2k^2}{a + b}.$$

(We may dismiss the case when $\mathbf{K}$ and $\mathbf{C}$ are concentric, for then the inverse of O is in infinity.)

The x-coordinate of the centre of the inverse is

$$c' = \frac{1}{2}(a' + b') = \frac{1}{2}\left(\frac{k^2}{a} + \frac{k^2}{b}\right).$$

The centre of the inverse coincides with the inverse of the centre if and only if $c' = \sigma'$, hence

$$\frac{2k^2}{a+b} = \frac{1}{2}\left(\frac{k^2}{a} + \frac{k^2}{b}\right).$$

This occurs if and only if $(a + b)^2 = 4ab$ or $(a - b)^2 = 0$, hence $a = b$. This means that the circle $\mathbf{C}$ *becomes a single point*. Hence if the radius of a circle is different from zero, the centre of its inverse cannot coincide with the inverse of its centre.

8. Let $\mathcal{H}$ be the hyperbolic family determined by $\mathbf{C}_1$ and $\mathbf{C}_2$ and $\mathcal{E}$ the elliptic family orthogonal to $\mathcal{H}$. Let S and T be the two points common to all the circles in $\mathcal{E}$. Let $\mathbf{K}$ be any circle with centre S. Then the inverse about $\mathbf{K}$ of the family $\mathcal{E}$ is a pencil of lines $\mathcal{L}$ intersecting at T', the inverse of T about $\mathbf{K}$.

Since the inversion preserves orthogonality, it follows that the inverses of the circles belonging to $\mathcal{H}$ are orthogonal to the lines belonging to $\mathcal{L}$, hence they are concentric circles, with their common centre at T'. In particular, $\mathbf{C}'_1$ and $\mathbf{C}'_2$, the inverses of $\mathbf{C}_1$ and $\mathbf{C}_2$ about $\mathbf{K}$, are concentric.

9. By Problem 8, the circles $\mathbf{C}_1$ and $\mathbf{C}_2$ can be inverted so that the inverses, $\mathbf{C}'_1$ and $\mathbf{C}'_2$ are concentric. Let r_1 and r_2 be the radii of $\mathbf{C}'_1$ and $\mathbf{C}'_2$. (Note that $r_1 > r_2$.)

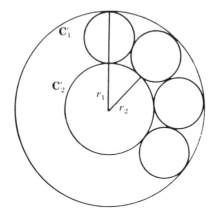

Figure 10.

If the ratio $\dfrac{r_1}{r_2}$ is such that it is possible to draw a chain of congruent circles, each touching its two neighbours and touching $\mathbf{C}'_1$ and $\mathbf{C}'_2$, then the number of circles in the chain is independent of their positions. (See Figure 10.) So the inverses of the circle chain belonging to $\mathbf{C}'_1$ and $\mathbf{C}'_2$ form a closed circle chain tangential to $\mathbf{C}_1$ and $\mathbf{C}_2$.

The number of circles is the same as in the inverse chain, though varying positions of the circles touching $\mathbf{C}'_1$ and $\mathbf{C}'_2$ give rise to different (not congruent) chains touching $\mathbf{C}_1$ and $\mathbf{C}_2$.

10.   The inverse of ℓ is the circle ℓ' going through S. Consider the elliptic family $\mathcal{E}$ of circles through A and B. Their centres lie on ℓ and they are orthogonal to ℓ. (See Figure 11.)

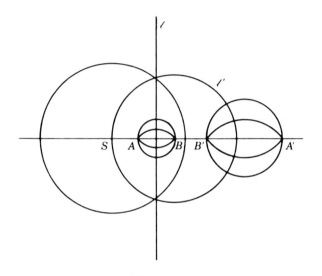

Figure 11.

The inversion about $\mathbf{K}$ taking A to A' and B to B' takes the family $\mathcal{E}$ to another elliptic family $\mathcal{E}'$ intersecting at A' and B' and orthogonal to ℓ'. It follows *that A' and B' are inverse with respect to ℓ'*.

Chapter 3

1.   Since the required "line" $\mathbf{p}$ is orthogonal to $\mathbf{L}$, it is generally a circle intersecting $\mathbf{L}$ orthogonally at A. Hence its centre X lies on the tangent line to $\mathbf{L}$ at A. X must also be on the perpendicular bisector b of OA, since O and A are points of the circle $\mathbf{p}$. Note that OX is *touching* $\mathbf{L}$ at O. See Figure 12.

Hence the construction gives a unique circle $\mathbf{p}$ as long as the perpendicular bisector b of OA intersects the tangent-line t to $\mathbf{L}$ drawn at O.

Exceptional case: b is parallel to t. This happens if and only if OA is perpendicular to t, hence OA is a diameter of $\mathbf{L}$. In this case OA is the required orthogonal p. See Figure 13.

Note also that $\mathbf{L}$ could be any line (in ordinary Euclidean geometry) going

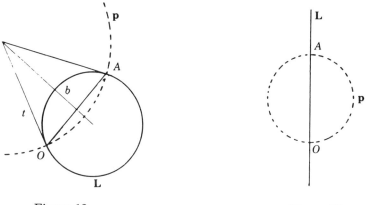

Figure 12.   Figure 13.

through O. In this case **p** is a circle of diameter OA.

Hence in D.W.E.G. it is *always* possible to find a line perpendicular to a given line at a given point, and the perpendicular line is *unique*.

2. Recall that if the inverse of A with respect to **L** is the point A', then any circle through A and A' is orthogonal to **L**. Thus A' must be found first and the required "line" (circle **p**) is the circle through the points O, A, and A'. Its centre is the intersection of the perpendicular bisectors of the segments OA, AA', respectively. The circle **p** exists and is unique unless O, A, and A' are collinear.

Exceptional cases:

a. O, A, and A' are on the same line. In this case the centre of **L** lies on the line OA. As in Problem 1, the *line OA* gives the *required orthogonal* **p**.

b. **L** is a line through O. Then **L** contains the centre of the required circle **p**, hence the centre of **p** is the intersection of **L** and the perpendicular bisector of OA.

Thus, in D.W.E.G., it is always possible to drop a perpendicular from a point to a given "line" **L** and the perpendicular is unique.

3. Let $\mathbf{a}, \mathbf{b}, \mathbf{c}$ be the three given "lines" (circles), intersecting pairwise: **a** and **b** at O and C; **b** and **c** at O and A; **c** and **a** at O and B. (See Figure 14.)

Denote the interior angles of the D.W.E.G. triangle by α, β, γ at the vertices A, B, C, respectively. These are enclosed by the tangent lines drawn at A, B, and C.

By symmetry the angles α', β', γ' enclosed pairwise by the tangent lines drawn at O to **a**, **b**, and **c** are such that

$$\alpha' = \alpha, \qquad \beta' = \beta, \qquad \gamma' = \gamma.$$

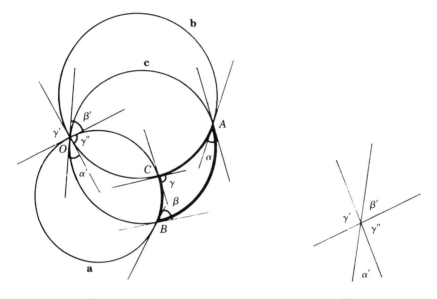

Figure 14.   Figure 15.

Their position is exhibited in Figure 15. Denote by γ'' the angle vertically opposite to γ' at O. Then

$$\alpha + \beta + \gamma = \alpha' + \beta' + \gamma' = \alpha' + \beta' + \gamma'' = 180°.$$

4. By Problem 2 of Chapter 1, a circle is orthogonal to every circle in a family of circles intersecting in two points (elliptic family), if its centre is on the radical axis (common chord) of the family.

Assume first that A is not on the line OP. Then there is a unique circle **S** through O, P, and A. The tangent t to **S** at the point A intersects OP at X. The circle of centre X and radius XA is orthogonal to **S** hence it is orthogonal to all the circles through O and P.

If A is on the line OP, let **S** be a circle with diameter OP. Let A' be the inverse of A with respect to **S**. The circle about diameter AA' has its centre on the line OP and is orthogonal to **S**, hence it is also orthogonal to the elliptic family intersecting at O and P.

If A is the midpoint of OP, then the perpendicular bisector line of OP is the required orthogonal.

5. a. By Problem 9 of Chapter 1 all circles orthogonal to a given circle **L** and going through a fixed point A intersect in A', the inverse of A about **L**. In particular, **C** goes through A'. (See Figure 16.)

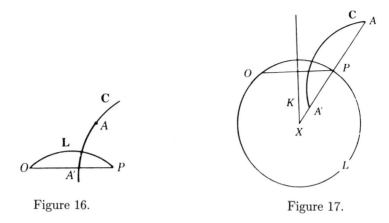

Figure 16.   Figure 17.

b. Let A and A' be two points on **C**. The centre X of the required circle **L** is the intersection of the line AA' and b, the perpendicular bisector of OP. (See Figure 17.)

Since by construction **C** is orthogonal to all the circles going through O and P, **C** is orthogonal to **L** (the circle with centre X and radius XP), and hence A' (the intersection of C and the line joining A to the centre X) is indeed the inverse of A in **L**.

Exceptional cases:

1. If AA' is parallel to b, then A' must be the reflection of A in OP. In this case **L** is the line OP. (See Figure 18.)

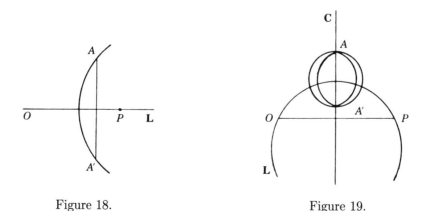

Figure 18.   Figure 19.

2. If **C** is the perpendicular bisector of OP (hence orthogonal to all the circles intersecting at O and P), then all circles through A and A' must be orthogonal to **L** (if A and A' are inverse in **L**). See Figure 19. So **L** is a circle through O

orthogonal to the elliptic family intersecting at A and A', hence can be found as in Problem 4.

The point P is the "centre" in D.W.E.G. of $\mathbf{C}$ but it is not the Euclidean centre of $\mathbf{C}$. (See Figure 20.)

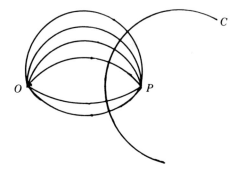

Figure 20.

6. Let $\mathbf{C}$ be a circle through A, touching $\mathbf{L}$ at O. Since $\mathbf{C}$ and $\mathbf{L}$ have a common tangent line t through O, the required construction consists of finding a circle through A and touching t at O. The centre X must be on the perpendicular bisector b of the segment AO and on a line p perpendicular to t through O. (See Figure 21.) Hence there is a unique solution if b is not parallel to p.

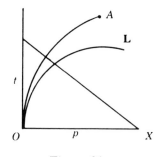

Figure 21.

Exceptional case: b is parallel to p, hence AO is perpendicular to p. This occurs if A lies on t. In this case t is a line in D.W.E.G. *parallel* to $\mathbf{L}$ (since O is not a point in D.W.E.G.). Thus there is a unique solution in every case.

7. Let $\mathbf{C}_1$ and $\mathbf{C}_2$ be two circles belonging to the families $\mathcal{P}_1$ and $\mathcal{P}_2$, respectively. Their intersections are T and Q. The tangents to $\mathbf{C}_1$ and $\mathbf{C}_2$ at T are t and u, respectively, where t and u are mutually perpendicular.

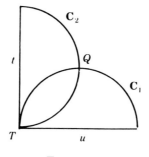

Figure 22.

The angle enclosed by the tangents to the two circles at Q is equal to the angle of the tangents at T, hence it is also a right angle. (See Figure 22.)

8.

a. Let t be the tangent line through O common to **L** and **M**. Let u be a line perpendicular to t at O. The circle through O, A, and C is orthogonal to **L**. Therefore, the tangent at O to the circle OAC is perpendicular to t, thus it is the line u. So the circle OAC belongs to the parabolic family orthogonal to the parabolic family containing the circles **L** and **M** (see Figure 23). Hence OAC is orthogonal to **M**. Similarly, the circle ODB is orthogonal to both **L** and **M**. (OAC and ODB are "parallel" lines in D.W.E.G.)

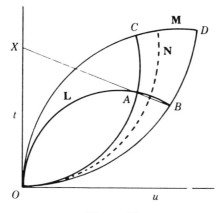

Figure 23.

b. The required circle **N** intersects the arcs AB and CD orthogonally, if it belongs to the parabolic family of **L** and **M**, hence its centre is on t. If B and A are inverse about **N**, then the centre of **N** is on the line

AB. Hence *X*, the centre of **N**, is the intersection of *AB* and *t* and the radius of **N** is the segment *OX*. The circle **M** is the inverse of **L** in **N**, and in particular *D* on **M** is the inverse of *C* on **L**, since the arc *CD* is orthogonal to **N**.

(*Note.* Alternatively, *X* can be determined as the intersection of the lines *AB* and *CD*.)æ

Chapter 4

1. Comparing triangles *QPX* and *PQY*, we have $PX = QY$ by construction, so

$$\triangle QPX \equiv \triangle PQY \qquad (SAS).$$

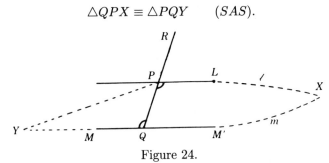

Figure 24.

Thus $\angle QPY = \angle PQX = 180° - \angle YQP$, since by assumption *X* lies also on *m*. (See Figure 24.)

Since $\angle XPQ = \angle YQP$, we have next

$$\angle QPY = 180° - \angle XPQ,$$

from which it follows that *Y* is on line ℓ. Hence ℓ and *m* intersect in *X and Y*, two points on opposite sides of *PQ*. This contradicts the axiom that two lines intersect in *at most one point*. Now, if the corresponding angles *RPL* and *PQM′* are known to be equal, then it follows that the supplementary angles *LPQ* and *MQP*, respectively, are also equal to each other and so it follows from the previous result that ℓ and *m* are parallel.

2. The diagram (Figure 25) consists of *n* copies of the original triangle $A_1 B_1 B_2$, with their bases on line ℓ. The assumption to be disproved is that the angles of $\triangle A_1 B_1 B_2$ have a sum greater than $180°$.

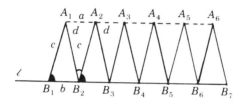

Figure 25.

i. By construction $\angle A_2 B_2 B_3 = \angle A_1 B_1 B_2$. The sum $\angle B_3 B_2 A_2 + \angle A_2 B_2 A_1 + \angle B_1 B_2 A_1 = 180°$. Hence it follows from the assumption that

$$\angle A_1 B_2 A_2 < \angle B_1 A_1 B_2.$$

It follows next from c(iv) of the introductory notes [p. 76] that $a < b$ (∗), (by comparing triangles $B_1 A_1 B_2$ and $A_1 B_2 A_2$).

ii. From the triangle inequality c(i) [p. 76] it follows first that the length of the path between two points is minimal if the path is along the line joining the two points. On the diagram the distance $B_1 B_6$ is less than the sum

$$B_1 A_1 + A_1 A_2 + A_2 A_3 + A_3 A_4 + A_4 A_5 + A_5 B_6.$$

In general, if the number of copies drawn is n, (hence there are $n + 1$ triangles), we have

$$c + na + d > (n + 1)b. \qquad (1)$$

We can make n as large as we wish. Dividing through by n in (1), we obtain

$$\frac{c}{n} + a + \frac{d}{n} > b + \frac{b}{n}.$$

When n is sufficiently large, $\frac{c}{n}, \frac{d}{n}, \frac{b}{n}$ can be made negligibly small. We then obtain $a > b$, which contradicts (∗) in (i).

3.

i. Since $\triangle PAB$ is isosceles, $\angle APB = \angle ABP$. See Figure 26. (This follows from congruence arguments, valid also in hyperbolic geometry.) The angle sum in triangle PAB does not exceed $180°$ (see Problem 2), hence

$$2\angle PBA + (180° - \angle PAQ) \le 180°,$$

whence $2\angle PBA \le \angle PAQ$.

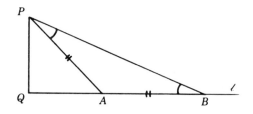

Figure 26.

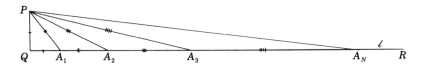

Figure 27.

ii. In Figure 27 $PQ = QA_1 = A_1A_2$, $PA_2 = A_2A_3$, $PA_3 = A_3A_4$, and so on.

Since the angle sum in $\triangle PQA$ does not exceed $180°$ and the triangle is isosceles,
$$\angle PA_1Q \leq 45°.$$

By (i),
$$\angle PA_2Q \leq 45°/2, \qquad \angle PA_3Q \leq 45°/4,$$
and so on with $\angle PA_nQ \leq (1/2)^{n-1}45°$.

Furthermore, if R is a point on ℓ, further from Q than A_n, then by c(ii) [see p. 76], $\angle PRQ \leq \angle PA_nQ$.

iii. Let m be perpendicular to PQ, so m is parallel to ℓ. (See Figure 28.) Assuming that there are at least two lines through P parallel to ℓ, let n be another parallel, enclosing angle α with m. By the result in (ii), the point R can be selected on ℓ so that
$$\psi = \angle PRQ < \alpha,$$

while $\beta = \angle RPQ < 90° - \alpha$, since n does not intersect ℓ, but PR does. Hence the sum of the angles in $\triangle RPQ$ (see Figure 29) is
$$90° + \beta + \psi < 90° + (90° - \alpha) + \alpha = 180°.$$

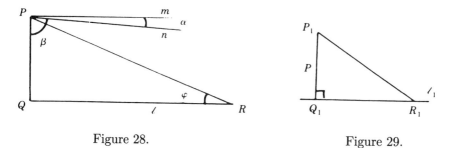

Figure 28.   Figure 29.

4.

i. Let ℓ_1 and ℓ_2 be two different lines in the hyperbolic plane. Choose P_1 and P_2 so that their respective perpendicular distances from ℓ_1 and ℓ_2 are equal. Let p be this common distance.

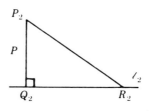

Figure 30.

Let Q_1 and Q_2 be the feet of the perpendiculars from P_1 and P_2 to ℓ_1 and ℓ_2, respectively, and choose R_1 on ℓ_1 and R_2 on ℓ_2 such that

$$Q_1 R_1 = Q_2 R_2.$$

Then $\triangle P_1 R_1 Q_1 \equiv \triangle P_2 R_2 Q_2$ (SAS). Hence $\angle R_1 P_1 Q_1 = \angle R_2 P_2 Q_2$. Let R_1 and R_2 vary so that the distances $Q_1 R_1$ and $Q_2 R_2$ increase, while remaining equal with each other. The limiting value of the angle $\angle R_1 P_1 Q_1$ is the same as that of the angle $R_2 P_2 Q_2$.

This limiting value is the angle of parallelism.

ii. The angular defect of every triangle in the hyperbolic plane is greater than 0. (This follows from Problem 3 and the arguments in the text.)

Let R be any point on ℓ distinct from Q. Let m be perpendicular to QP at P and let n enclose an angle $\alpha = \angle PRQ$ with PR. (See Figure 31.) Then, by the converse of the parallel axiom, both m and n are parallel to ℓ and they are distinct, since $\alpha < 90° - \angle RPQ$ (because the angular defect in $\triangle PQR$ is positive). So the angle of parallelism, measured between the critical parallel and PQ, is not greater than the angle between n and PQ, hence less than 90°.

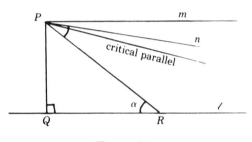

Figure 31.

iii. Let P' be on QP. Let S be a point on the critical parallel through P
so that $\phi = \angle SPQ$ is the angle of parallelism (at P). (See Figure 32.)

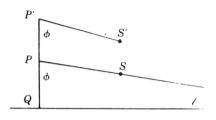

Figure 32.

Through P' draw a ray $P'S'$ so that the angle $S'P'Q = \phi$. Then $S'P'$
is *parallel to PS* (by the converse of the parallel axiom). The line PS
divides the plane into two half-planes. All points of ℓ are in one half,
since ℓ does not intersect PS. Since $PS, P'S'$ do not intersect, all the
points of $P'S'$ are in the other half-plane. So $P'S'$ does not intersect ℓ,
hence the critical parallel ray to ℓ through P' cannot enclose a greater
angle with $P'Q$ than ϕ.

5.

i. Let α, β, γ be the angles of the triangle. Partition by cutting α into α_1
and α_2. Let the angles of the part triangles be α_1, β, ϕ_1 and α_2, γ, ϕ_2,
respectively.

The angular defects of the part-triangles are $\alpha_1 + \beta + \phi_1 - 180°$ and
$\alpha_2 + \gamma + \phi_2 - 180°$, respectively.

Their sum is

$$\alpha_1 + \alpha_2 + \beta + \gamma + \phi_1 + \phi_2 - 360° = \alpha + \beta + \gamma - 180°.$$

This is the defect of the original triangle.

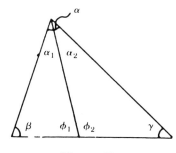

Figure 33.

ii. Referring to the figure given, observe that triangles PQ_1T_1 and $Q_2Q_1T_1$ are congruent (SAS), hence the angular defect in each is the same. Hence by the additivity of defects, the defect in $\triangle PT_1Q_2$ is twice the defect in $\triangle PQ_1T_1$. It follows that the *defect in $\triangle PQ_2T_2$ is more than twice* the defect in $\triangle PQ_1T_1$. This repeats in every step, so the angular defect in triangle PQ_nT_n is more than 2^{n-1} times the defect in $\triangle PQ_1T_1$. However, the defect cannot be equal to or more than $180°$, and so in a finite number of steps a certain point Q_n will be reached such that the perpendicular at Q_n to PQ_n cannot intersect PT_1T_2, hence the ray PT_1T_2 is parallel to the perpendicular to PQ_n at Q_n. So the angle of parallelism is α if PT_1 is the *critical* parallel (to the perpendicular to PQ_n drawn at Q_n), but in the general case the critical parallel differs from PT_1 and then the critical angle is less than α (arbitrarily chosen). Hence the angle of parallelism is at least α, if Q_n is far enough, and from Problem 4, it follows that if $p > PQ_n$, then the angle of parallelism cannot become greater.

6.

i. Join AE, DE, and EF. Then $\triangle ABE \cong \triangle DCE$ (SAS), hence $DE = AE$, $\angle EDC = \angle EAB = \alpha$, and $\angle DEC = \angle AEB = \gamma$.

Next $\triangle AFE \cong \triangle DFE$ (SSS), hence $\angle AFE = \angle DFE = 90°$.

Also $\angle DEF = \angle AEF = \delta$.

Thus $\delta + \gamma = \angle BEF = \angle CEF = 90°$ and so FE is perpendicular to both AD and BC. It follows that AD and BC are parallel.

ii. Since triangles AFE and DFE are congruent, it follows also that

$$\angle FAE = \angle FDE = \beta.$$

Hence $\angle BAF = \angle CDF = \alpha + \beta.$

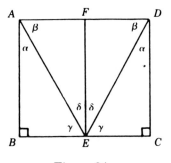

Figure 34.

The sum of the angles of the quadrilateral $BCDA$ is less than 360° (the quadrilateral can be partitioned into two triangles, each having angles with sum less than 180°). Since $ABCD$ has right angles at B and C the sum of the equal angles at A and D is less than 180°, so each of the two angles is acute.

iii. Draw the diagonal BD. (See Figure 35.)

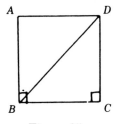

Figure 35.

In $ABCD$ the sum of the angles at B and D is less than 90° since $\angle DCB = 90°$. Since $\angle CBD + \angle DBA = 90°$, it follows that $\angle ABD > \angle BDC$. The triangles ABD and BDC have two equal sides enclosing unequal angles. By c(iv) (of the introductory notes on p. 76) $AD > BC$.

7.   Let Q and Q' be two points on m, on the same side of M, Q' further removed. Construct a Saccheri quadrilateral with base LP, right angles at L and P, and $LM = PN$.

As seen in Problem 6, the angles at M and N are then equal *acute* angles. Since by the condition given the angle QML is 90°, it follows that MN is in the *interior* of the angle QML and it intersects PQ *between* P and Q. Thus $PN < PQ$, where by construction $PN = LM$, so $LM < PQ$ as claimed. The quadrilateral $PLMQ$ has three right angles, hence the fourth angle at Q must be acute. So the exterior angle $Q'QP$ is obtuse. Construct another Saccheri quadrilateral on the base PP'

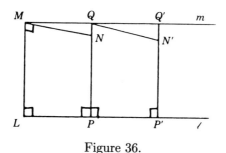

Figure 36.

with $P'N' = PQ$. Then the angle PQN' is acute and so it is the interior of the obtuse angle $Q'QP$, so QN' intersects $P'Q'$ internally. Hence, similarly as before

$$P'N' < P'Q' \quad \text{where} \quad P'N' = PQ.$$

8. Let the triangles ABC and $A'B'C'$ have equal angles at their vertices A and A', B and B', C and C' respectively. *Suppose* that the triangles are *not congruent*, hence they differ in at least one of their sides. Assume that $A'B' < AB$. (See Figure 37.)

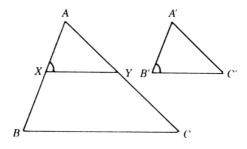

Figure 37.

Let X and Y be points on the lines AB, AC, respectively, such that

$$AX = A'B' \quad \text{and} \quad \angle AXY = \angle A'B'C'.$$

Then $\triangle AXY \cong \triangle A'B'C'$ (ASA), hence

$$\angle AYX = \angle A'C'B' = \angle ACB.$$

By construction $\angle AXY = \angle A'B'C' = \angle ABC$ and so $\angle YXB + \angle CBX = \angle YXB + \angle AXY = 180°$.

Furthermore, since $\angle AYX = \angle ACB$ as shown, it follows similarly that $\angle XYC + \angle BCY = 180°$. Thus the sum of the angles in the quadrilateral $BXYC$ is $360°$. This however cannot happen in the hyperbolic plane. Thus we have a contradiction.

9.

i. The area of the lune is $\frac{\alpha}{360} \cdot A$, where A is the surface area of the sphere.

ii. Denote the area of the spherical triangle PQN by T_0 and that of triangle PQS by T_1. Then, by (i)

$$T_0 + T_1 = \frac{\alpha}{360} A \tag{1}$$

Similarly, supplementary triangles of areas T_2 and T_3, respectively, complete PQN into lunes of angles β and γ, respectively, hence

$$T_0 + T_2 = \frac{\beta}{360} A \tag{2}$$

and

$$T_0 + T_3 = \frac{\gamma}{360} A. \tag{3}$$

The surface of the sphere is made up of eight triangles, each area T_0, T_1, T_2, T_3 occurring twice. Hence

$$T_0 + T_1 + T_2 + T_3 = \frac{A}{2}. \tag{4}$$

Using (1), (2), and (3), equation (4) becomes

$$T_0 + \frac{\alpha + \beta + \gamma}{360} A - 3T_0 = \frac{A}{2} \,,$$

whence

$$T_0 = \frac{\alpha + \beta + \gamma - 180}{180} \cdot \frac{A}{4} = \frac{\varepsilon}{180} \cdot \frac{A}{4} \,.$$

Chapter 5

1. This may be tedious because so many different cases have to be considered.

  Case (a)   Suppose that the points A, B, C lie in the same half-plane determined by the line ℓ. Their images are A', B', C' and AA', BB', CC' intersect ℓ in P, Q, R, respectively. (See Figure 38.)

  We have to show that the angles $A'B'Q$ and $C'B'Q$ are equal to ABQ and CBQ respectively. It follows then that $A'B'Q$ and $C'B'Q$ are supplementary angles since the angles ABQ and CBQ are supplementary and so $A'B'C'$ must lie on a line.

Proof of the equality of the angles. $\triangle APQ \cong \triangle A'PQ$ (SAS). Hence $AQ = A'Q$ and $\angle A'QP = \angle AQP$. It follows that $\angle A'QB' = \angle AQB$. Hence $\triangle A'QB' \cong \triangle AQB$ (SAS) and so $\angle A'B'Q = \angle ABQ$. By starting from triangles CRQ and $C'RQ$ we show similarly that $\angle QBC = \angle QB'C'$. This completes the proof.

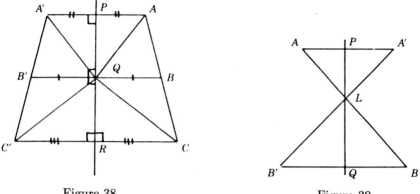

Figure 38.   Figure 39.

Case (b)   Suppose that the points A, B, C lie on a line intersecting ℓ at a point L, and at least two of them, say A and B, lie in different half-planes. (See Figure 39.)

We prove then that A', B', and L are on the same line by showing that the vertically opposite angles $A'LP$ and $B'LQ$ are equal.

$$\triangle APL \cong \triangle A'PL \text{ (SAS)} \quad \text{and} \quad \triangle BQL \cong \triangle B'QL \text{ (SAS)}$$

Hence $\angle A'LP = \angle ALP$ and $\angle B'LQ = \angle BLQ$. Since $\angle ALP = \angle BLQ$ as A, L, B are on a line, it follows that $\angle A'LP = \angle B'LQ$ as desired.

It follows now that the images of *all the points* of a line must be on the same line, by fixing two points of the given line and using the fact that the image of the third point arbitrarily chosen on the line lies on the line determined by the images of the first two points.

i. *Case (a)*   One of the lines forming the angle is ℓ. The vertex of the angle is P and m' is the reflected image of line m. Let A be on m and A' be the image of A, the intersection of AA' and ℓ being L.

   Then $\triangle PLA \cong \triangle PLA'$ (SAS) hence the reflected angle $A'PL = \angle APL$ as required, noting that the sense of rotation is changed.

ii. *Case (b)*   The vertex P is on line ℓ, but its two arms are different from ℓ. The arms m and n may be in the same half-plane or in different half-planes.

   By the result (a) the angles between ℓ and m' and ℓ and n' respectively are the same as those between ℓ and m and ℓ and n'. The angle between

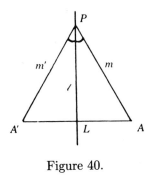

Figure 40.

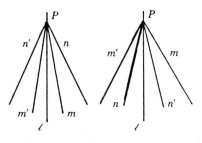

Figure 41.

m' and n' is the difference or the sum of the angles enclosed with line ℓ, hence it is equal to the difference (or sum) of the angles between m and ℓ and n and ℓ and so equal to the angle between m and n (different sense).

iii. *Case (c)* The vertex P is not on the line ℓ. Let Q be a point on ℓ. Draw a perpendicular at Q to ℓ intersecting m, n, m', n' at A, B, A', B', respectively. (See Figure 42.)

Clearly $QA = QA'$ and $QB = QB'$.

$$\triangle P'LQ \cong \triangle PLQ \text{ (SAS)},$$

therefore, $PQ = P'Q$ and $\angle LQP = \angle LQP'$, and thus

$$\angle PQA = \angle P'QA' = \angle P'QB' = \angle PQB.$$

From this $\triangle PQA \cong \triangle P'QA'$ (SAS) and so $\angle QPA = \angle QP'A$ and similarly $\triangle PQB \cong \triangle P'QB'$ hence $\angle QPB = \angle Q'P'B'$. By subtraction $\angle APB = \angle A'P'B'$ as required.

2. By property (d) the point P' lies on the perpendicular from P to ℓ. Let M be the intersection of PP' and ℓ and let L be a point on ℓ different from M. (See Figure 43.)

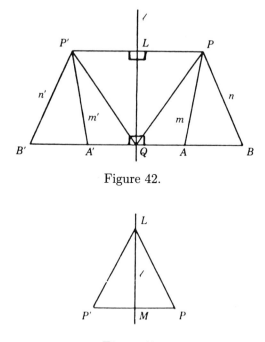

Figure 42.

Figure 43.

By property (e) $\angle MLP' = \angle MLP$. Thus $\triangle MLP \cong \triangle MLP'$ (ASA). Thus $MP' = MP$. This proves that P' is a reflection of P in ℓ. (Note that P' is not the same as P as the angles $P'LM$ and PLM are of opposite sense.)

3.

   i. *Case a.* Let one end of the line segment be on ℓ. Then the image of AB is AB'. Let BB' intersect ℓ in Q. From the congruence of the triangles AQB and AQB' (SAS) it follows immediately that $AB = AB'$. (See Figure 44.)

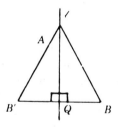

Figure 44.

Case b.

AB intersects ℓ at P, so A, B lie in different half-planes, A', B' are the images. From Case (a) it follows that $AP = A'P$ and $BP = B'P$. Hence $AB = A'B'$.

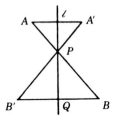

Figure 45.

Case c.

The general case. Let A, B lie in the same half-plane. Let A', B' be the images of A and B, and AA', BB' intersecting ℓ in P, Q respectively. $\triangle APQ \cong \triangle A'PQ$ (SAS) hence $A'Q = AQ$ and $\angle PQA' = \angle PQA$. Thus $\angle A'QB' = \angle AQB$.

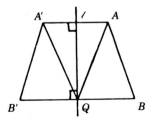

Figure 46.

It follows that $\triangle A'QB \cong \triangle AQB$ (SAS), hence $A'B' = AB$. It follows immediately that triangles reflect into congruent triangles since neither the measures of the sides nor those of the angles change by reflection.

ii. Suppose that the segments AB and $A'B'$ are congruent, and A, A' and B, B' are distinct. Then one reflection about ℓ, the perpendicular bisector of AA', will take A to A' and B to some point B_1. If B_1 is the same as B', then one reflection is sufficient. If B_1 differs from B', we have $A'B_1 = AB$ by (i) and since $AB = A'B'$, $A'B_1 = A'B'$. Let the line m bisect the angle $B'A'B_1$. Let M be the intersection of $B'B_1$ and m. Then $\triangle MA'B_1 \cong \triangle MA'B'$ (SAS) hence $MB' = MB$, and $\angle B'MA' = \angle B_1MA' = 90°$. Thus reflecting about m, A' remains the same and B' is the reflection of B_1.

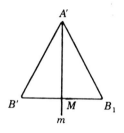

Figure 47.

Suppose now that the triangles ABC and $A'B'C'$ are congruent. Then AB is brought into $A'B'$ by at most two reflections as seen. (See Figure 48.) Suppose that these two reflections bring C into C_1, a position different from C'. Then $A'C_1 = A'C'$ and $B'C_1 = B'C'$, since the two reflections preserve the lengths of AC and BC and $AC = A'C'$, $BC = B'C'$. Thus $\triangle A'B'C_1 \cong \triangle A'B'C'$ (SSS) and in particular $\angle C'A'B' = \angle C_1A'B'$. Hence a reflection about $A'B'$ brings C_1 into C' as shown before.

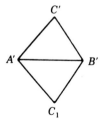

Figure 48.

4. We make use of the result that all circles orthogonal to a given circle **C** and passing through a point A, different from the centre of **C**, intersect at A', the inverse of A with respect to **C**.

At least one of the points A and B is different from the centre of **P**. Suppose that A is not the centre and its inverse is A'. Then the required circle must pass through A, A' and B. If the three points are not collinear, there is a unique circle circumscribed around A, A' and B. If they are collinear, then the required orthogonal to **P** is a diameter of **P**. This is clearly the case if B coincides with the centre of **P**. In the more general situation, if B is not the centre of **P**, B being collinear with A and A', lies on the diameter determined by the inverse pair A, A'. In any case, two distinct points determine a *unique* Poincaré line.

5.

i. Assume that $ST \parallel AB$. Then $SABT$ is a symmetric trapezium, since $\angle TAB = \angle STA$ (alternating) and $\angle STA = \angle SBA$ (subtending the same arc) and so $\angle TAB = \angle SBA$ so that these angles subtend equal chords AS and BT.

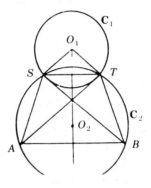

Figure 49.

In a symmetric trapezium the median (joining the centres of the parallel sides) is perpendicular to the parallel sides. (To see this, you may use the exercises on reflection.) Since ST is a chord of $\mathbf{C}_1$, this perpendicular bisector passes through O_1, the centre of $\mathbf{C}_1$. Since the perpendicular bisector is also the perpendicular bisector of AB, it follows that $O_1A = O_1B$.

Conversely, if $O_1A = O_1B$, then the central line O_1O_2 perpendicularly bisects AB, since $O_2A = O_2B$ also. The line O_1O_2 also perpendicularly bisects ST, the common chord of the two circles. Hence AB and ST are perpendicular to the same line and so they are parallel.

ii. If B is the inverse of A in ℓ, then all circles through A and B intersect ℓ orthogonally (Chapter 1, Problem 9).

Let S be any point on $\mathbf{P}$. Construct a circle $\mathbf{C}$ through A, B, S to intersect $\mathbf{P}$ again in T. Then ST is the common chord (the radical axis) of two circles $\mathbf{P}$ and $\mathbf{C}$, both orthogonal to ℓ. Hence the centre of ℓ must be on the radical axis of $\mathbf{P}$ and $\mathbf{C}$, i.e., on ST. It must also lie on AB, since an inverse pair of points is collinear with the centre of the circle of inversion. Hence the intersection of AB and ST determines uniquely the centre of ℓ as Q. A tangent from the point Q to $\mathbf{P}$ gives the radius of ℓ. (See Figure 50.)

This construction fails if and only if ST is parallel to AB. Using the result in (i), this happens if and only if $OA = OB$. In this case ℓ is a

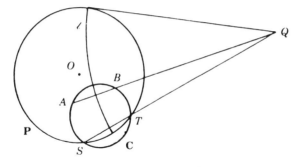

Figure 50.

line, the angular bisector of the angle AOB and the inversion becomes a reflection in the ordinary sense.

6.

i. Let **C** be the given circle, with centre O, and P the given point on line t. The required circle **S** is supposed to touch t at P, so its centre is on a line ℓ perpendicular to t at P.

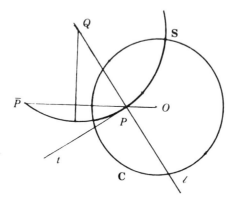

Figure 51.

Since the circle **S** is to be orthogonal to **C** it must also pass through $\overline{P}$, the inverse of P in **C**. Thus the centre of **S** is the intersection point Q of ℓ and the perpendicular bisector of $P\overline{P}$. The radius of **S** is given by QP. The solution is unique.

Exceptional case: ℓ and the perpendicular bisector of $P\overline{P}$ are parallel. In this case t is the line $P\overline{P}$, a line through the centre 0. The required circle now becomes a line, namely the line t, which is orthogonal to **C**, passes through P, and is "tangential" to itself.

ii. Draw tangents to ℓ and m at A and bisect the angles between them. Denote the two angular bisectors (perpendicular to each other) by t and t'. The required angular bisector circles are then circles orthogonal to **P** and touching t and t', respectively, at A. (See Figure 52.) The construction in (i) gives the required circles. (One of these can turn out to be a diameter of **P**.)

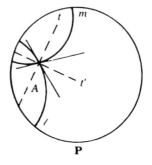

Figure 52.

iii. Constructing on m segments PQ and PR congruent in the Poincaré sense to AB :

Step 1: Find an axis of inversion (see Problem 5) to invert A into P. (See Figure 53.) Suppose that $\overline{B}$, the inverse of B by this construction, also falls on line m. Then $\overline{B}$ can be taken to be Q, since $P\overline{B}$ is then a segment of the Poincaré line m and is congruent to AB as required.

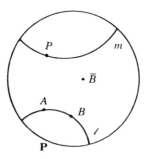

Figure 53.

In this case, step 2 consists of inverting $\overline{B} = Q$ in a line through P orthogonal to m. (The line is a circle orthogonal to **P** and m and passing through P, which can be found as before (see Problem 12, Chapter 1). The inverse of Q is R, and PR is congruent to PQ, hence to AB.)

The general case: $\overline{B}$ is not on m. Then step 2 is using the construction in (ii) to find the angular bisector circles of m and the arc $P\overline{B}$. Step 3 is then the inversion of $\overline{B}$ into Q and R about the angular bisector circles. Then PQ and PR are congruent to $P\overline{B}$, hence to AB.

7.

i. By definition of Poincaré congruence, $A'B'$ is the image of AB after successive inversions in Poincaré lines. These inversions bring the line ℓ to line ℓ', hence the image of C is also on ℓ'. BC is known to be congruent to $B'C'$ and there are only two points on ℓ' on either side of B satisfying this condition. Since *betweenness* is preserved in inversion, C' must be the image. This shows that $A'C'$ is congruent to AC.

ii. It is to be shown that if the given conditions are satisfied, then a number of inversions in Poincaré lines brings $\triangle ABC$ into $\triangle A'B'C'$.

Since $A'B'$ is congruent to AB, a composition of inversions can be found such that $A'B'$ is the image of AB. Let the image of C under the same composition of inversions be $\overline{C}$. Since inversions preserve angles, the angle between $A'B'$ and $A'\overline{C}$ is equal to α and, therefore, to α', the angle between $A'B'$ and $A'C'$. One arm, namely $A'B'$, of these two angles is common. If the other arm is also common, then $\overline{C}$ coincides with C' since AC and $A'C'$ are known to be congruent. If the other arm does not coincide, then an inversion about $A'B'$ leaves $A'B'$ fixed and brings $A'\overline{C}$ into $A'C$ since the angles are equal. Hence the Poincaré triangle ABC transforms into $A'B'C'$ in a finite number of inversions about Poincaré lines.

Note. As in Euclidean geometry, the other congruency cases follow as theorems from the SAS case.

8.

i. Since $\angle ADB$ is exterior to triangle ADM it is greater than the nonadjacent interior angle AMD. Hence $\angle AMD$ is acute. In triangles AMB and AMC: side AM is common; $BM = MC$; while angle AMB is acute; hence $\angle AMC$ is obtuse. Thus the side AC (opposite the larger angle) is greater than side AB.

ii. Let R be the intersection of the line PQ and the circle **C**. Then $\angle ARP$ is acute since one of its arms is the diameter of circle **C**. Furthermore $RQ > RP$.

Proof. Let r be the radius of circle **C**. Let $RP = x$, $RQ = y$. Then $(r - x)(r + y) = r^2$ since Q is the inverse of P. Hence $y(r - x) = rx$ or $y = \frac{r}{r-x} \cdot x$, where $r - x < r$.

Thus by (i), side AQ of $\triangle APQ$ is longer than side AP, so arc $AQ >$ arc AP.

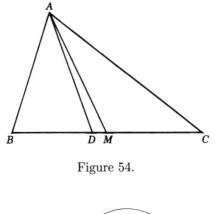

Figure 54.

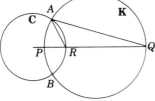

Figure 55.

iii. We treat the figure as a Euclidean figure; Poincaré-congruent segments generally have different Euclidean lengths. Let d be the diameter of circle **P**, orthogonal to ℓ and intersecting ℓ in M. (See the figure on the left below.) We show first that if $B_{i-1}B_i$ and B_iB_{i+1} are two adjacent segments on ℓ (on the *same side of M*) and are Poincaré congruent and such that B_{i+1} is closer to M than B_i, then the arc B_iB_{i+1} is of greater Euclidean length than arc $B_{i-1}B_i$.

Proof. Let $\mathbf{C}_i$ be the Poincaré line through B_i orthogonal to ℓ (that is a circular arc orthogonal to **P** and ℓ through B_i). Then, by the definition of Poincaré congruence, B_{i-1} and B_{i+1} are inverse about $\mathbf{C}_i$, B_{i-1} being internal and B_{i+1} external to $\mathbf{C}_i$. It follows from (ii) that the arc B_iB_{i+1} is longer than arc $B_{i-1}B_i$.

We may now assume that the points A_1, A_2, and P are on the same side of M, for if not, then some point A_k in the sequence (identifying the sequence with the B_iB_{i+1}, treated above; at some stage A_k will get on the other side of M) is on the same side as P, so we may take this point as a starter.

Define the sequence of *distinct* points $\{C_0, C_1, C_2, \ldots, \}$ in the following manner: $C_0 = P$, C_1 is *between* A_1 and P and the segments $C_0C_1, C_1C_2 \ldots$ are Poincaré congruent to A_1A_2.

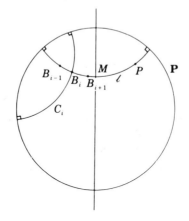

Figure 56.

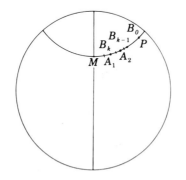

Figure 57.

It follows from (ii) that the Euclidean lengths of the segments $C_0C_1, C_1C_2, C_2C_3, \ldots$ form an *increasing* sequence. Thus for some k, A_2 is between A_1 and C_{k-1}, while C_k is between M and A_2, or coincides with A_2.

In the latter case, it follows from the Poincaré congruence of the segments involved that $A_2 = C_k$, $A_3 = C_{k-1}$, and so on. Thus $C_0 = P = A_{k+2}$ and so P lies between A_1 and A_{k+3}. Hence $N = k+3$ satisfies the requirement of the theorem.

If $C_k \neq A_2$, then C_k *is between* A_1 *and* A_2, since the segments A_1A_2 and $C_{k-1}C_k$ are Poincaré congruent and so A_1 and A_2 cannot both be between C_{k-1} and C_k. It follows similarly that C_{k-1} is between A_2 and A_3, so the points of the A and C sequences alternate.

So $C_0 = P$ is between A_{k+1} and A_{k+2}. $n = k+2$ satisfies the requirement.

9.

i. The required Poincaré line n is an arc of a circle through Q, orthogonal to ℓ and **P**, intersecting **P** in A. (See Figure 58.) For the construction see Chapter 1, Problem 8.

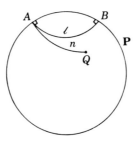

Figure 58.

ii. Invert L into the point C as shown in Problem 5(ii), the circle of inversion being **K**, orthogonal to both ℓ and **P**. Hence each ℓ and **P** invert into themselves about **K**, (exchanging arcs internal and external to **K**). So the intersection A of ℓ and **P** goes into the other intersection B. QL inverts into an arc orthogonal to ℓ and **P** and since this arc passes through C, the centre of AB, it becomes the diameter OCS' (O being the centre of **P**).

The inverse of Q in **K** is the point P on OC, so by definition CP is Poincaré congruent to QL. The inverse of the arc QA is then the arc PB and since inversion keeps the angles unchanged, the angle between the arcs QA and QL (by (i) the angle of parallelism) goes into the angle

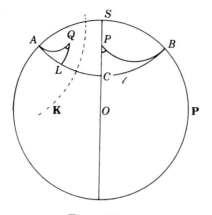

Figure 59.

between PC and the arc PB. (Alternatively, the same angle is enclosed by PC and the arc PA orthogonal at A to **P**.)   Either of these two angles measures the angle of parallelism at P with respect to ℓ and is equal to the angle of parallelism at Q.

iii. The angle of parallelism at P is the angle between CP and the tangent t at P to the Poincaré line PA. Let t intersect the radius OA at R. Since the arc PA is orthogonal to **P**, the radius OA is tangential to arc AP.

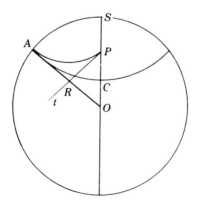

Figure 60.

So the line segments RP and RA are equal. (Tangents from the external point R.)   Hence in the isosceles triangle ARP, the angles RAP and APR are equal.

$$\angle OPR = \angle APO - \angle APR = \angle APO - \angle RAP = \angle APO - \angle OAP.$$

Clearly, if P' is a point between S and P then $\angle AP'O < \angle APO$ since the angle APO is exterior to triangle $AP'P$. (See Figure 61.)

Since $\angle OAP' > \angle OAP$ it follows that $\angle OP'R < \angle OPR$, i.e., the angle of parallelism decreases as P moves from C to S.

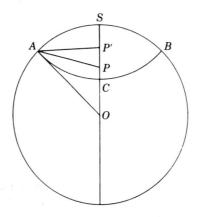

Figure 61.

The limiting values are:

(i) The angle of parallelism when P coincides with C, in which case the critical parallel coincides with ℓ, hence the angle between OC and ℓ is $90°$.

(ii) When P reaches S, we have $\angle ASO - \angle OAS = 0$ since the triangle ASO is isosceles. (In this case the circle orthogonal to $\mathbf{P}$ at A and going through S is orthogonal to $\mathbf{P}$ at S, hence the tangent to the circle at S coincides with OS.)

10.   Let A be a point on $\mathbf{K}$. Consider the set of points arising from inverting A in the circles belonging to the family $\mathcal{E}$.

Since $\mathbf{K}$ is orthogonal to any circle $\mathbf{S}$ of the family $\mathcal{E}$, the inverse of A in $\mathbf{S}$ is again on $\mathbf{K}$, while the inverse of C in $\mathbf{S}$ is again C (since C is a point of S). Conversely, C remains invariant through inversion, if it is on the circle of inversion, while the Poincaré segments CA and CA' are Poincaré congruent only if A and A' are inverse about a Poincaré line, hence an arc of a circle orthogonal to $\mathbf{P}$ and so belonging to $\mathcal{E}$. Therefore, for given Poincaré points C and A, the Poincaré circle $\mathbf{K}$ is *uniquely* defined as the circle belonging to the family $\mathcal{H}$, going through A. (A cannot belong to more than one circle in $\mathcal{H}$.)

If C is at the (Euclidean) centre of $\mathbf{P}$, then the elliptic family of circles becomes the family of lines through C, while the associated hyperbolic family becomes the set of concentric circles of (Euclidean) centre C. Hence the Euclidean circles with centre C and inside $\mathbf{P}$ are also Poincaré circles.

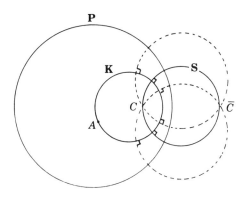

Figure 62.

11.

i. Let k be the radius of the circle of inversion. Then inversion leads to a pair of circles of radii a' and b' where

$$a'a = b'b = k^2,$$

hence

$$\frac{a'}{b'} = \frac{b}{a},$$

so

$$\log a' - \log b' = \log b - \log a = -(\log a - \log b).$$

A second inversion about another concentric circle would result in a pair of circles of radii a'', b'' with

$$\log a'' - \log b'' = \log b' - \log a' = \log a - \log b.$$

In each case, the absolute value of $\log a - \log b$ is unchanged, hence $d = |\log a - \log b|$ is invariant.

For the three circles of radii a, b, c, we have, by definition, inversive distances:

$$d_1 = |\log a - \log b|, \qquad d_2 = |\log b - \log c|, \qquad d_3 = |\log a - \log c|.$$

If $a > b > c$, then all three differences are positive, hence the $\|$ sign may be omitted and we have $d_1 + d_2 = d_3$.

If $a < b < c$, then all three differences are negative, thus

$$d_1 = \log b - \log a, \qquad d_2 = \log c - \log b, \qquad d_3 = \log c - \log a.$$

We obtain again $d_1 + d_2 = d_3$.

ii. The inverse of **P** is a line through E_2, perpendicular to E_1E_2, hence, with the segments internal and external to E_1 and E_2 changing into each other, E_2 inverts into itself. An arbitrary circle through E_1 and E_2 inverts into a line through E_2.

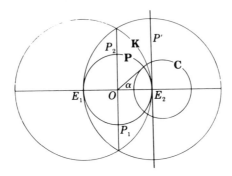

Figure 63.

The inverse of the line P_1P_2 is a circle, orthogonal to E_1E_2, through E_1, hence its radius is $2r$. The inverse of **C** is a circle of centre E_2 and radius a_1 calculated as follows.

Let S be the centre of **C**, and A_1 the inverse of A in **K**. Then

$$E_1A = E_1O + OA = E_1O + OS - SA = r + r\sec\alpha - r\tan\alpha. \quad (1)$$

Then the radius of the inverse of **C** is

$$a_1 = E_2A_1 = E_1A_1 - E_1E_2, \quad (2)$$

where
$$E_1A_1 = (2r)^2/E_1A = (2r)^2/r(1 + \sec\alpha - \tan\alpha) \quad (3)$$

and
$$E_1E_2 = 2r.$$

Substituting (1) and (3) into (2) and simplifying we obtain

$$a_1 = 2r\,\frac{\cos\alpha + \sin\alpha - 1}{\cos\alpha - \sin\alpha + 1}.$$

Putting $\cos\alpha = \dfrac{1-t^2}{1+t^2}$ and $\sin\alpha = \dfrac{2t}{1+t^2}$ we obtain $a_1 = 2rt$ where $t = \tan\dfrac{\alpha}{2}$.

iii. Since inversion preserves Poincaré congruence, we may consider the segments A_1B_1 and C_1D_1 which are the inverses of AB and CD in **K**.

Let $\mathbf{C}_A = \mathbf{C}, \mathbf{C}_B, \mathbf{C}_C, \mathbf{C}_D$ be circles orthogonal to $\mathbf{P}$ and intersecting OE_2 orthogonally at A, B, C, D. By (ii), they invert to concentric circles of centre E_2 and of radii $2r \tan \frac{\alpha}{2}$, $2r \tan \frac{\beta}{2}$, $2r \tan \frac{\gamma}{2}$, $2r \tan \frac{\delta}{2}$, where β, γ, δ are defined similarly to α. The segments $A_1 B_1$ and $C_1 D_1$ are Poincaré congruent if they invert into each other, that is, if the circle-pairs of radii $2r \tan \frac{\alpha}{2}$, $2r \tan \frac{\beta}{2}$ and $2r \tan \frac{\gamma}{2}$, $2r \tan \frac{\delta}{2}$ invert into each other. In that case, it follows from (i) that

$$\left| \log \tan \frac{\alpha}{2} - \log \tan \frac{\beta}{2} \right| = \left| \log \tan \frac{\gamma}{2} - \log \tan \frac{\delta}{2} \right|.$$

Since $(OA) = \log \cot \frac{\alpha}{2} = -\log \tan \frac{\alpha}{2}$, and similar formulae hold for (OB), (OC), and (OD), we obtain from above that

$$|(OA) - (OB)| = |(OC) - (OD)|$$

or

$$(AB) = (CD).$$

Using the result in (i) for *inversive* distances,

$$d_3 = d_1 + d_2,$$

we obtain

$$\left| \log \tan \frac{\alpha}{2} - \log \tan \frac{\gamma}{2} \right| = \left| \log \tan \frac{\alpha}{2} - \log \tan \frac{\beta}{2} \right| + \left| \log \tan \frac{\beta}{2} - \log \tan \frac{\gamma}{2} \right|.$$

Introducing $\cot \frac{\alpha}{2}$, $\cot \frac{\beta}{2}$, $\cot \frac{\gamma}{2}$ again, we have

$$(AC) = (AB) + (BC).$$

Chapter 6

1.

a. A figure B is said to be *congruent* to figure A if a succession of reflections in lines transforms A into B. It suffices therefore to show that if B is the reflection of A in a line m, then A and B are projectively equivalent.

Let O be any point on m, P a point belonging to figure A, P' its image on B. Then the distances OP and OP' are equal. Produce the segment PO to P'' such that the distances PO and OP'' are equal. Let d be the perpendicular distance of P to m, equal to the distance of P' to m.

Then P'' is again at *distance d* from m and on *the side of m as* P'. Hence P' and P'' are on a line parallel to m. This construction can be carried out for all points: $P_1, P_2, P_3, \ldots$ of A, keeping O *fixed*.

Then the figures $P_1'' P_2'' P_3'', \ldots$ and $P_1 P_2 P_3, \ldots$ are in perspective from O, while the lines $P_1' P_1'', P_2' P_2'', \ldots$ determine lines parallel to m. Thus a perspective transformation with vertex O, followed by a parallel perspective transformation, takes the figure A to figure B. This proves the claim.

Note that the figures A and B need not be coplanar. For each point P of A the line m and P determine a plane containing the points O, P', P'', hence $P'P'' \parallel m$ as shown.

For $P_1, P_2, \ldots$ on A, the lines $P_1' P_1'', P_2' P_2''$ are all parallel to m (though not necessarily coplanar), so in every case $P_1' P_2' \ldots$ is a parallel projection of $P_1'' P_2'' \ldots$.

b. Consider first the case of two triangles $A_1 B_1 C_1$ and $A_2 B_2 C_2$ such that

$$A_1 B_1 \| A_2 B_2, \qquad B_1 C_1 \| B_2 C_2, \qquad C_1 A_1 \| C_2 A_2.$$

The two triangles are equiangular and, therefore, similar (congruent if $A_1 B_1 = A_2 B_2$).

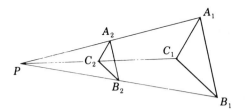

Figure 64.

The lines $A_1 A_2, B_1 B_2, C_1 C_2$ are *concurrent*, for if P is the intersection of $A_1 A_2$ and $B_1 B_2$, then

$$PB_1/PB_2 = PA_1/PA_2 = A_1 B_1/A_2 B_2,$$

and if Q is the intersection of $B_1 B_2$ and $C_1 C_2$, then

$$QB_1/QB_2 = QC_1/QC_2 = B_1 C_1/B_2 C_2 = A_1 B_1/A_2 B_2.$$

Hence $PB_1/PB_2 = QB_1/QB_2$ and so $Q = P$. So the two triangles are in perspective from P and hence projectively equivalent. (See Figure 64.)

Note. the two triangles need not be coplanar.

Next, consider the general case of two similar triangles $A_1B_1C_1$ and $A_2B_2C_2$. We construct a triangle $A_3B_3C_3$ with sides parallel to those of $A_1B_1C_1$ and $A_3B_3 = A_2B_2$, so that $A_3B_3C_3$ and $A_2B_2C_2$ are congruent. Then by (a) the triangles $A_3B_3C_3$ and $A_2B_2C_2$ are in perspective from some point. Hence $A_1B_1C_1$ and $A_2B_2C_2$ are projectively equivalent.

Note. While the proofs in (a) and (b) establish that congruence and similarity are special cases of projectivity, they do not provide constructions for carrying out the transformations in the smallest possible number of steps.

2. Assume first that $ABCD$ is not a trapezium. Then the sides AB, DC intersect in P and the sides AD, BC intersect in Q. Make the line PQ the *vanishing* line of the projective transformation. Construct a plane $\mathcal{P}_2$ intersecting $\mathcal{P}_1$ in some line j parallel to PQ, and choose V in some plane $\mathcal{P}_3$ intersecting $\mathcal{P}_1$ in PQ.

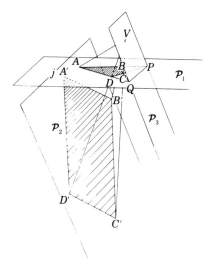

Figure 65.

Then the planes $\mathcal{P}_2$ and $\mathcal{P}_3$ are parallel, thus the images of P and Q in $\mathcal{P}_2$ are in infinity. So the images of AB and DC are parallel lines $A'B', D'C'$ and similarly $A'D'$ and $B'C'$ are parallel.

If $ABCD$ is a trapezium, where $AB\|DC$, then BC and AD intersect in P.

This time choose for a vanishing line the line through P parallel to AB and DC and find $\mathcal{P}_2$ and V the same way as before. Then the images $B'C'$ and $A'D'$ are parallel (by the same reasoning as before) while the planes VAB and VCD intersect $\mathcal{P}_3$ in lines parallel to j (therefore parallel to AB and DC) hence $A'B'$ and $D'C'$ are also parallel.

3.

a. Let $ABCD$ be a parallelogram in plane $\mathcal{P}$. Construct a square $AB'C'D$, such that AB' and DC' are perpendicular to $\mathcal{P}$ and $AB' = DC' = AD$. Join BB' and CC'. Then $BCC'B'$ is a parallelogram, since $B'C'$ is parallel to AD, hence to BC and $B'C' = AD = BC$. So CC' and BB' are parallel. Draw also lines through A, D parallel to BB'. Then the square $ADC'B'$ and the parallelogram $ADCB$ are in parallel perspective.

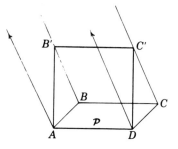

Figure 66.

b. First, we show that all squares are projectively equivalent to a fixed square S. Let T be any square. Construct a square S' with sides equal to those of S and parallel to the sides of T. Then T and S' are in perspective from a point, while S and S' being congruent are projectively equivalent. Then T and S' are *projectively equivalent* from a point, while S and S' being congruent are projectively equivalent. Hence S and T are projectively equivalent.

From (a) it follows that all parallelograms are projectively equivalent to squares, while from Problem (2) it follows that all plane quadrangles are projectively equivalent to parallelograms. Finally, a skew quadrangle (that is a quadrangle $ABCD$ whose vertices are not in the same plane) can be brought into perspectivity with some plane quadrangle $ABCD'$, choosing some vertex V not in the plane of ABC and finding D' as the intersection of VD and the plane ABC.

Thus all quadrangles are projectively equivalent.

4. The lines VAA', VBB' determine a plane, hence the lines $AB, A'B'$ intersect in some point Z. Similarly, $BC, B'C'$ intersect in X, and $AC, A'C'$ in Y. (See Figure 67.) As AB lies in the plane $\mathcal{P}_1$, so does Z, since Z is a point on the line AB. Since Z is also a point of $A'B'$, Z is in plane $\mathcal{P}_2$. *Thus Z is on the intersection line ℓ of $\mathcal{P}_1$ and $\mathcal{P}_2$.* The same thing applies to X and Y, hence X, Y, Z are collinear.

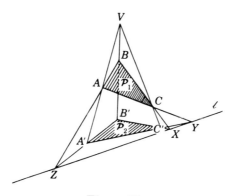

Figure 67.

Conversely, if X, Y, Z are known to lie on the same line ℓ (in fact, it suffices to know that X, Y, Z *exist*, it follows then that they are on the intersection line ℓ of $\mathcal{P}_1$ and $\mathcal{P}_2$) it follows that $AA'BB'$, $BB'CC'$, $CC'AA'$, respectively, determine three planes. V is the common point of these planes and V lies on each of the three intersection lines AA', BB', CC'. Hence ABC, $A'B'C'$ are in perspective from V.

Note. The cases of parallel perspectivity, or the parallelness of the planes $ABC, A'B'C'$, or of a pair of sides (e.g., $AB, A'B'$) are not regarded as special cases in projective geometry: intersection points of lines could be ideal.

5. (1) Desargues' theorem is proved in Problem 4 for ABC and $A_1B_1C_1$ since they are not coplanar. Thus the line pairs $\{AB, A_1B_1\}$, $\{BC, B_1C_1\}$, and $\{AC, A_1C_1\}$ intersect on the line a. Since the intersections of the lines BC, CA, and AB are known to be X, Y, Z, respectively, these same points are also intersections of a with B_1C_1, C_1A_1, and A_1B_1, respectively, and so the lines B_1C_1, C_1A_1, and A_1B_1 also intersect $B'C'$, $C'A'$, and $A'B'$ in X, Y, Z.

Since $A_1B_1C_1$ and $A'B'C'$ are not in the same plane, the theorem can be applied again: the triangles $A'B'C'$ and $A_1B_1C_1$ are in perspective from some point V'.

(2) Denote the planes A_1VV', B_1VV' and C_1VV' by Π_A, Π_B, Π_C, respectively. The line AA' is in Π_A, since A is on VA_1 and A' is on $V'A_1$. Hence AA' and VV' are in the same plane and, therefore, intersect in some point. Similarly BB' and CC' are in Π_B and Π_C, respectively, and thus intersect VV'. Since AA', BB' and CC' are in plane $\mathcal{P}$, their intersections with VV' must coincide with the intersection of VV' with $\mathcal{P}$, which is a point W. Thus ABC and $A'B'C'$ are in perspective from W.

(3) Keeping notations as before, assume that the triangles ABC and $A'B'C$ (now in the same plane) are in perspective from a point V and that the pairs $\{BC, B'C'\}$, $\{CA, C'A'\}$, $\{AB, A'B'\}$ intersect in X, Y, Z, respectively. We have to show that X, Y, Z are collinear.

We apply the converse theorem, proved in (2), to the triangles $AA'Y$ and $BB'X$, knowing that AA' and BB' intersect in V, AY and BX intersect in C, and $A'Y$ and $B'X$ intersect in C' where V, C, C' are collinear.

So the triangles $AA'Y$ and $BB'X$ are in perspective. The centre of perspectivity is the intersection of AB and $A'B'$, hence it is the point Z. Thus the third line XY also passes through Z, which means that X, Y, Z are collinear as claimed.

6. In the Desargues configuration we may take $BCC'B'$ to be a parallelogram, since any quadrilateral can be transformed through a sequence of perspectivities into a parallelogram, each perspectivity preserving collinearity of points and concurrence of lines.

The special case is shown in Figure 68.

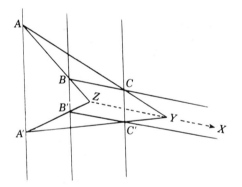

Figure 68.

Since the centre of perspectivity of triangles ABC and $A'B'C'$ is infinity, it follows that AA' is also parallel to BB' and CC'.

Since BC and $B'C'$ are parallel, their intersection point X is in infinity. To show that Z and Y, the intersections of the pairs $\{AB, A'B'\}$ and $\{AC, A'C'\}$ are collinear with X, it need be shown that ZY is parallel to BC and $B'C'$.

In triangles $B'BZ$ and $A'AZ$ we have

$$BZ/AZ = BB'/AA' \tag{1}$$

and in triangles $C'CY$ and $A'AY$

$$CY/AY = CC'/AA'. \tag{2}$$

Since the distances CC' and BB' represent opposite sides of a parallelogram,

$$CC' = BB'$$

and so from (1) and (2)

$$BZ/AZ = CY/AY, \quad \text{hence} \quad AB/AZ = AC/AY. \tag{3}$$

It follows from (3) that BC is parallel to ZY as required.

7. *Case 1.* Let ℓ_1 and ℓ_2 intersect in O.
Triangles OA_1B_2 and OA_2B_1 are similar, thus

$$\frac{OB_2}{OA_2} = \frac{OA_1}{OB_1} \tag{1}$$

From similar triangles OA_1C_2 and OC_1A_2

$$\frac{OC_2}{OA_2} = \frac{OA_1}{OC_1} \tag{2}$$

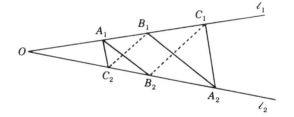

Figure 69.

From (1) and (2) it follows that

$$\frac{OB_2}{OC_2} = \frac{OC_1}{OB_1}.$$

Hence B_1C_2 is parallel to B_2C_1. (See Figure 69.)

Case 2. Now there are two parallelograms: $A_1B_1A_2B_2$ and $A_1C_1A_2C_2$. Hence the segments A_1B_1 and B_2A_2 are equal, and segments A_1C_1 and C_2A_2 are equal.

Thus $A_1C_1 - A_1B_1 = C_2A_2 - A_2B_2$ or $B_1C_1 = B_2C_2$. So $B_1C_1B_2C_2$ is also a parallelogram, hence $B_1C_2 \| C_1B_2$. (See Figure 70.)

The general case for the Pappus configuration is as shown in Figure 71.

The intersections of A_1B_2 with B_1A_2 and of A_1C_2 with C_2A_1 are C_3 and B_3 respectively.

The intersection of B_1C_2 and B_2C_1 is A_3. Denote by ℓ_3 the line B_3C_3. To show that A_3 is on ℓ_3 make ℓ_3 the vanishing line of a pespectivity. To do this draw

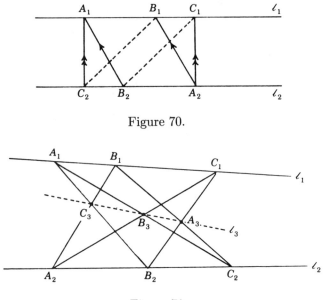

Figure 70.

Figure 71.

parallel planes $\mathbf{P}_1$ and $\mathbf{P}_2$ intersecting the plane $\mathbf{P}$ of the figure in ℓ_3 and in ℓ', respectively, where ℓ' is parallel to ℓ_3 and selecting V as the centre of perspectivity in $\mathbf{P}_1$.

In the projection, ℓ_3 is the ideal line of $\mathbf{P}_2$. Hence the projections of A_1B_2 and B_1A_2 are parallel and so are the projections of A_1C_2 and A_2C_1.

Case (1) arises when the intersection of ℓ_1 and ℓ_2 is not on ℓ_3. Then its projection does not vanish, so the images of ℓ_1 and ℓ_2 are not parallel. Otherwise the intersection point of ℓ_1 and ℓ_2 is ideal, so the projections are parallel. This leads to case (2).

In either case it was shown that the projections of B_1C_2 and C_1B_2 intersect on the ideal line, the image of B_3C_3, hence A_3 is on B_3C_3.

8. Let ℓ be the line (think of a "club") containing points ("members") $P_1, P_2, \ldots, P_k$.

i. By axiom 3 there exist in the plane 4 points such that no three lie on a line. Thus the plane must contain *at least two more points Q_1 and Q_2*, not on ℓ.

By axiom 1, each pair of points, $(Q_1, P_1), (Q_1, P_2), \ldots, (Q_1, P_k)$, determines a unique line. Let these lines be $\ell_1, \ell_2, \ldots, \ell_k$, respectively. These lines are *distinct*. (Suppose for example that ℓ_1 and ℓ_2 coincide. Then P_1, P_2, Q_1 are on the same line. But there is *only one* line, namely ℓ, containing P_1 and P_2 and by assumption Q_1 is not on ℓ.)

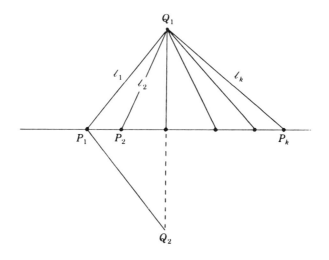

Figure 72.

Furthermore we show now that we can find at least one line m in the plane, distinct from $\ell, \ell_1, \ell_2, \ldots, \ell_k$. Amongst the lines $\ell_1, \ell_2, \ldots, \ell_k$ find one which does not contain Q_2. (Since Q_1 is the unique intersection point of any two of these lines, Q_2 lies on at most one of the lines.) The points $P_1, P_2, \ldots, P_k$ can be labelled so that ℓ_1 is a line not containing Q_2. Then $m = Q_2 P_1$ is a line distinct from ℓ_1, since ℓ_1 does not contain Q_2, and is distinct from all the other lines $\ell_2, \ldots, \ell_k$, since those do not contain P_1, and is also distinct from ℓ, since Q_2 is not on ℓ.

We show next that the line m *contains exactly k points* and, more generally, that any line not through Q_1 contains exactly k points. (This is true for ℓ by assumption.) By axiom 2, each of $\ell_1, \ldots, \ell_k$ intersects m in a point, which is unique by axiom 1. Furthermore, these intersection points are distinct, since $\ell_1, \ldots, \ell_k$ have only Q_1 as a common point. Thus m contains *at least k points*. Interchanging the roles of ℓ and m, we can also see that ℓ contains at least as many points as m, hence m *contains exactly k points*, and this applies by the same argument for all lines not through Q_1.

To show that each of the lines $\ell_1, \ldots, \ell_k$ also contains exactly k points, we can apply the same arguments: replacing the point Q_2 by Q_1 and considering the intersections of the lines $Q_2 P_1, \ldots, Q_2 P_k$ with ℓ and $\ell_1, \ldots, \ell_k$, excepting the line $Q_1 Q_2$. To complete the proof, take any point P not on $Q_1 Q_2$ and different from P (such a point must exist, otherwise axion 3 is not satisfied). Join P with the k points of ℓ_1. Together these points intersect $Q_1 Q_2$ in k points and the previous arguments can be applied. This completes the proof of (i). The rest follows now easily.

ii. Through each point P there are k distinct lines obtained by joining P with the k points of some line not through P. There are no more lines through P, since by axiom 2 each line intersects any other line in some of its k points.

iii. Fix a point P. All the points of the plane lie on the k lines through P since by axiom 1, any point P', distinct from P, determines a line PP' through P.

Since there are k lines through P, each of them containing $k - 1$ points other than P, the total number of points in the plane is

$$k(k - 1) + 1 = k^2 - k + 1.$$

iv. Fix a line ℓ. By axiom 2, each line of the plane intersects ℓ in one of its k points. Since through each of the k points there are $k - 1$ lines other than ℓ, it follows that the total number of lines in the plane is

$$k(k - 1) + 1 = k^2 - k + 1.$$

This completes the proof.

Note. (a) Observe that in axioms 1 and 2 the words "point" and "line" are interchangeable if you also interchange the words "through" and "on." This is the principle of duality. In the problem just discussed, (ii) is the *dual* of (i) and (iv) the dual of (iii).

(b) It is clear from the solution of (i) and axiom 3 that k is at least 3, so the Fano plane represented in the "gemmetry" of the Introduction and the schedule of the meetings of the friends of Alice is the smallest possible finite projective geometry, with 7 points, 7 lines, 3 points on each line, 3 lines through each point.

Section IIII

On Books to Follow up the "Journey"

When you return from a successful tour, you feel that there are places where you would have liked to dwell longer than the short tour allowed you. On finishing this somewhat whimsical journey into fascinating, albeit not expansive, realms of modern geometry, here are some suggestions for helping to broaden and deepen your experience.

The first and most suitable book dealing with our special subject, appealing because of its historical background, its comfortable, informal style, yet careful and rigorous attention to detail, is

Euclidean and Non-Euclidean Geometries Development and History, 2nd ed., Marvin Jay Greenberg, W. H. Freeman and Company, San Francisco, 1979.

This book also gives a detailed bibliography for further reading. However, the following shorter list represents books dealing with broad aspects of geometry and which you can read without excessive mathematical apparatus. These books are (in date order):

Forder, H. G., *Geometry,* Hutchinson, London, 1950.

Hilbert, D. and S. Cohn-Vossen, *Geometry and the Imagination,* Chelsea, New York, 1952.

Eves, H., *A Survey of Geometry,* Allyn and Bacon, Boston, 1963–65.

Coxeter, H. S. M., *Introduction to Geometry,* Wiley, New York, 1969.

Pedoe, D., *A Course of Geometry,* Cambridge University Press, New York, 1970.

Coxeter, H. S. M. and S. L. Greitzer, *Geometry Revisited,* Mathematical Association of America, Washington D.C., 1975 .

In addition to the books on general geometry listed above, here are three more books, each completing some particular aspect touched upon in the "Journey."

The classic on the history of non-Euclidean geometry is:
Bonola, R., *Non-Euclidean Geometry,* Dover, New York, 1955.

For a detailed and easily readable exposition of the logical foundations of modern geometry see:
Tuller, A., *Modern Introduction to Geometries,* Van Nostrand Reinhold, New York, 1967.

For filling in the gaps in the discussion of the axiom-systems of Euclidean, absolute, and Non-Euclidean geometries look at:
Forder, H. G., *The Foundations of Euclidean Geometry,* Dover, New York, 1958.